Mindfulness in Behavioral Health

Series Editor: Nirbhay N. Singh

For further volumes:
http://www.springer.com/series/8678

Alexander Moreira-Almeida
Franklin Santana Santos

Editors

Exploring Frontiers of the Mind-Brain Relationship

 Springer

Editors
Alexander Moreira-Almeida
School of Medicine
Federal University of Juiz de Fora - UFJF
36036-330 - Juiz de Fora - MG
Brazil
alex.ma@ufjf.edu.br

Franklin Santana Santos
Institute of Psychiatry
School of Medicine
São Paulo University
05403-010 - São Paulo - SP
Brazil
franklin@saudeeducacao.com.br

ISBN 978-1-4614-0646-4   e-ISBN 978-1-4614-0647-1
DOI 10.1007/978-1-4614-0647-1
Springer New York Dordrecht Heidelberg London

Library of Congress Control Number: 2011940687

Printed on acid-free paper

Springer is part of Springer Science+Business Media (www.springer.com)

*To Angélica, Caio, and Laura, gifts of life
who give meaning to my life and reasons
to try to be a better human being every
single day.*

*To my parents, Hélio and Elizabeth,
who gave me not only the gift of life,
but have also been models of life,
and unlimited sources of support
and love.*

A.M.A.

*To my beloved parents, Dalva and Fernando,
who have taught me about Life and Love.*

F.S.S.

Foreword

Human beings are naturally curious and seek a coherent understanding of life and its mysteries. The possibility of life after death is the most ancient of all mysteries, as shown by the burial rituals of the earliest *Homo sapiens* (Sussman and Cloninger 2011). In fact, wonder about spirituality emerged along with the development of narrative language, art, and science. Human art, science, and spirituality are all basic expressions of the self-aware consciousness that is unique to modern human beings (Cloninger 2004, 2007; Sussman and Cloninger 2011). Hence, a fundamental characteristic of human beings is our need to explore and understand the frontiers of the relationships among body, thought, and spirituality.

Human beings have been aptly described as "evolution conscious of itself" in the insightful words of the anthropologist Sir Julian Huxley (1959). Accordingly, it is natural for human beings to try to understand the nature of the cosmos and their place within it in order to know how to satisfy their needs and live well in health and happiness. In order to live our whole life well, it is necessary to recognize that the conscious experience of human beings includes learning how to adapt in a wide range of circumstances. In fact, human beings need to adapt to five major types of situations: sexual, material, emotional, intellectual, and spiritual (Cloninger 2004). Among these five adaptive situations, spirituality is the most recently evolved and its evolution may be incomplete, thereby resulting in marked differences between different people in traits like altruism and gifts like clairvoyance or extrasensory perception. Spirituality is defined as the search for what is beyond human existence (Cloninger 2007). Direct personal experience of the transcendent is a part of most people's lives. Most people have had peak experiences of inseparability or oceanic feelings regardless of their religious beliefs or doubts (Cloninger 2004; Hay 2007). For example, most people report that they "sometimes have felt like I was part of something with no limits or boundaries in time and space" or "often feel so connected to the people around me that it is like there is no separation between us." Furthermore, people spend more time in prayer or meditation than they do having sex (Cloninger 2004; Hamer 2004). The fact that such self-transcendent experiences are such a frequent and inspiring aspect of human life suggests that science can never understand human nature well without investigating self-transcendent phenomena. Reconnecting science and spirituality is important for having a rational and comprehensive understanding of humanity and the world (Walach and Rech 2005).

Human spiritual needs have raised perennial questions about how to understand near-death experiences and the possibilities of life after death, such as reincarnation, wandering spirits of the dead, and states of spiritual possession or mediumship. These basic spiritual questions have great implications for our outlook on life, so it is not surprising that the suggested answers have led to much speculation and controversy. For example, Freud expressed doubt in any belief in an afterlife because it could be explained as seeking satisfaction from a wishful fantasy. However, his skepticism about human spirituality was based on a logical error (Cloninger 2007). Just because we desire something does not mean it is untrue or wishful fantasy. People often desire food to satisfy their material hunger, but that does not mean that food is not real. The desires and needs of human beings exist because they serve a real function. The maturity of a human being requires integration of the full range of their sexual, material, emotional, intellectual, and spiritual needs in a coherent reality-based manner (Cloninger 2006).

What people believe and the assumptions they make about life and human nature are highly dependent on cultural influences. People who live in a single culture or who reject inquiry into other cultures can have a difficult time recognizing the narrow influence of their particular culture on their thinking. Actually, the materialistic assumptions that are dominant in modern Western cultures are atypical of other modern human cultures. Around the world belief in wandering spirits and reincarnation are commonplace and not associated with any evidence of wishful fantasy (Stevenson 1983). Such facts do not tell us that such spiritual phenomena are true, but only that different cultures make different assumptions. As a result, a scientific person needs to establish reliable facts about spiritual phenomena and to test alternative ways of explaining the facts.

Some of our greatest scientists have been intensely preoccupied with understanding spiritual phenomena, including Newton and Wallace. At the end of the nineteenth century, there was widespread interest in spiritual phenomena among academics until many supposed mediums were exposed as frauds (Kottler 1974). Since then there has been great resistance among academics to even consider the possible reality of life after death as suggested by mediums and clairvoyants. Only the most courageous of empirical scientists like Hans Eysenck (Eysenck and Sargent 1993) and Ian Stevenson (1983) have been outspoken about their findings supporting paranormal abilities like extrasensory perception or recall of past lives. Although there are notable exceptions like Francis Collins (2006), most leading scientists today are highly materialistic and reject belief in anything transcendent, such as belief in God or life after death (Larson and Witham 1998). As a result, there is great social pressure among scientists to reduce all scientific explanations to material mechanisms and to reject consideration of any phenomena that cannot be explained by materialistic mechanisms as an impossible foolishness or the result of inadequate scientific rigor in observation. Despite persistent pressure from materialists (Larson and Witham 1998), an increasing majority of people in the general population have a spiritual awareness of something beyond human existence, even if they are not religious (Hay 2007). Whereas nearly 75% of the academic elite reject transcendent phenomena, more than 85% of the general population accepts them (Larson and Witham 1998; Ecklund and Long 2011). Interestingly, the rejection of transcendent phenomena by academics is related to their social background, rather than their intelligence and expertise (Ecklund and Long 2011; Evans and Evans 2008).

In any culture, the healthiest and happiest individuals have a "creative" personality configuration characterized by high development of Self-directedness, Cooperativeness, and Self-transcendence as measured by the Temperament and Character Inventory (Cloninger 2004; Cloninger and Zohar 2011). Despite this, longitudinal studies in secular cultures show that Self-directedness and Cooperativeness increase, but Self-transcendence decreases between adolescence and age 45. Then between 45 and 65 years of age people increase in Self-transcendence again as they learn to cope with suffering and death (Cloninger 2003). That is, the influence of secular culture often works to reduce Self-transcendence up to middle age, even though self-transcendence is characterized by being more unified in one's perspective on life and happier for all configurations of other personality traits (Cloninger and Zohar 2011). In the general population, poor development of Self-transcendence is characterized by unhappiness, reduced self-worth, and feelings of emptiness and alienation from other people and the world as a whole. Neglect of transcendent phenomena in science is also likely to have a high cost, particularly in efforts to reduce consciousness to the zombie-like state of physical mechanisms alone in which subjectivity, self-directedness, and free will are regarded as illusory (Cloninger et al. 1993).

Fortunately, substantive progress is being made at a philosophical and empirical level to redress the explanatory gap between physical and subjective accounts of consciousness in the understanding of mind-body relationships. The courageous academic contributors to this book point out that his Highness, The Materialistic Emperor, is wearing no clothes – that is, the promise of reductionistic materialism to explain everything has been made repeatedly but without success.

Nevertheless, the burden on alternative paradigms is to show the greater utility and explanatory power of more general models that allow for the three components of human beings – body, thoughts ("mind 1"), and psyche ("mind 2") (Cloninger 2004). In Chalmer's terminology, "mind 1" refers to intellectual reasoning based on semantic learning, whereas "mind 2" refers to creativity, free will, giftedness, and other self-transcendent abilities like extrasensory perception (ESP) (Chalmers 1996). The greatest problem of alternative models has always been the fact that paranormal phenomena often involve veridical functions along with wishful fantasy and/or fraud. Of course it is a logical error to conclude nothing is real because some examples are not, but how can scientists exclude the noise of specious claims by some people who falsely proclaim paranormal abilities?

When I was developing a measure of spirituality called the Self-transcendence scale as a component of the Temperament and Character inventory (Cloninger et al. 1993), I was chagrined to find that belief in paranormal abilities like ESP was a reliable indicator of high Self-transcendence. The belief in ESP is as characteristic of

Self-transcendence as are peak-experiences of boundlessness and inseparability (Cloninger 2004). I considered just eliminating such paranormal items to avoid criticism from materialists, but chose to respect the truth and reliability of my findings by describing the phenomena I was observing in an open-minded way. I am glad that I did so because otherwise I might have overlooked or misunderstood some clinically important phenomena about the expression of spirituality.

High Self-transcendence is characterized by creativity and wisdom when it is combined with high Self-directedness, but it is characterized by magical thinking and perceptual aberrations when combined with low Self-directedness (Cloninger 2004; Smith et al. 2008). In other words, appreciation of the wonders and mysteries of life always promotes good feelings, but some thoughts that make a person feel good can be wishful self-deceptions. Consequently, paranormal experiences can be produced by either healthy extraversion or unhealthy psychoticism using the terminology of Hans Eysenck (Eysenck and Sargent 1993). For people to enjoy realistic and productive lives, they must combine imaginative inquiry with rigorous reality testing, as do creative artists, scientists, and mystics. Likewise for reproducible results in science, people who report paranormal experiences need to be screened for the maturity and integration of their personality.

The dual nature of Self-transcendence is at the crux of the scientific challenge of studying spiritual phenomena about consciousness. There are genuine and reproducible transcendent phenomena to be understood, but there is also much superstition and deception. The insincere make it difficult and challenging to identify the sincere, but it is a severe error of logic to dismiss what is real because of some examples of fraud or fantasy. Some instances of spiritual phenomena are difficult to dismiss by an open-minded person, as documented throughout this courageous and informative book. Edgar Cayce, for example, is a particularly well-documented case of paranormal (i.e., transcendent) giftedness (Evans and Evans 2008). It is important to recognize that even such outstanding examples are not perfectly accurate, just as observers of real life events are not consistently precise.

It is useful to remember Plato's allegory of the cave in which most observers are like prisoners doomed to observing shadows of representations of reality, whereas only a few find their way to a direct and undistorted vision of reality (Plato 1977). This book on the exploration of the frontiers of the mind-body relationship brings together some precious observations about the fundamental mystery of the nature of consciousness. It is fascinating to read regardless of one's current beliefs about spirituality and mind-body phenomena. Its merit is not in the justification of definitive conclusions, but rather the opposite. It raises many questions that serve to invite each of us to be more aware of the uncertainty of our preconceptions about consciousness. Fortunately, the reader can rest assured that speculative reason can be disciplined by scientific rigor to specify testable questions about reproducible phenomena.

This book on the frontiers of mind-body relationships is a scholarly embodiment of creative and open-minded science. All open-minded people are clearly reminded that strict materialism is a specious and inadequate paradigm – the unhealthy and naked emperor of our scientific era. To restore balance to scientific inquiry, we need only recognize that the consciousness of human beings has a triune nature, one that has developed hierarchically over our long evolutionary history, including procedural learning of habits and skills in our early vertebrate ancestors, semantic learning of symbols and facts in anthropoid apes and early humans, and self-aware learning of narrative language, art, science, and spirituality in modern human beings (Sussman and Cloninger 2011; Cloninger 2009).

In my own opinion, we can best serve scientific truth by open-minded inquiry into the powerful interactions among material, cognitive, and spiritual mechanisms because the ternary components of consciousness never operate in isolation from one another (Cloninger and Cloninger 2011). A scientist cannot control what s/he does not measure or chooses to ignore by denial of its reality. In contrast, we can avoid the pitfalls of reductionism by using an integrative psychobiological approach, thereby staying alert to the full range of phenomena that can inform us about the triune nature of human consciousness.

St. Louis, MO, USA   C. Robert Cloninger, MD

References

Chalmers, D. J. (1996). *The conscious mind: In search of a fundamental theory*. New York: Oxford University Press.

Cloninger, C. R. (2003). Completing the psychobiological architecture of human personality development: Temperament, character, and coherence. In U. M. Staudinger & U. E. R. Lindenberger (Eds.), *Understanding human development: Dialogues with lifespan psychology* (pp. 159–182). London: Kluwer Academic Publishers.

Cloninger, C. R. (2004). *Feeling good: The science of well-being* (p. 374). New York: Oxford University Press.

Cloninger, C. R. (2006). The science of well-being: an integrated approach to mental health and its disorders. *World Psychiatry, 5*(2), 71–76.

Cloninger, C. R. (2007). Spirituality and the science of feeling good. *Southern Medical Journal, 100*(7), 740–743.

Cloninger, C. R. (2009). The evolution of human brain functions: The functional structure of human consciousness. *Australian and New Zealand Journal of Psychiatry, 43*(11), 994–1006.

Cloninger, C. R., & Cloninger, K. M. (2011). Development of instruments and evaluative procedures on contributors to Illness and health. *International Journal of Person-centered Medicine, 1*(1): 43–52.

Cloninger, C. R., Svrakic, D. M., & Przybeck, T. R. (1993). A psychobiological model of temperament and character. *Archives of General Psychiatry, 50*(12), 975–990.

Cloninger, C. R., & Zohar, A. H. (2011). Personality and the perception of health and happiness. *Journal of Affective Disorders, 128*(1–2), 24–32.

Collins, F. S. (2006). The language of god: A scientist presents evidence for belief. New York: Simon and Schuster.

Ecklund, E. H., & Long, E. (2011). Scientists and spirituality. Sociology of Religion. doi: 10.1093/socrel/srr003

Evans, J. H., & Evans, M. S. (2008). Religion and science: Beyond the epistemological conflict narrative. *Annual Review of Sociology, 34*, 87–105.

Eysenck, H. J., & Sargent, C. (1993). Explaining the unexplained: Mysteries of the paranormal. London: Prion.

Hamer, D. (2004). The god gene. New York: Doubleday.

Hay, D. (2007). Something there: The biology of the human spirit. Philadelphia: Templeton Press.

Huxley, J. (1959). Foreword. In P. T. D. Chardin (Ed.), The phenomenon of man (pp. 11–28). New York: Harper & Row.

Kottler, M. J. (1974). Alfred Russel Wallace, the origin of man, and spiritualism. Isis, 65(227), 145–192.

Larson, E. J., & Witham, L. (1998). Leading scientists still reject god. Nature, 394(6691), 331.

Plato. (1977). The republic of Plato (61st ed.). New York: Oxford University Press.

Smith, M. J., et al. (2008). Temperament and character as schizophrenia-related endophenotypes in non-psychotic siblings. Schizophr Research, 104(1–3), 198–205.

Stevenson, I. (1983). American children who claim to remember previous lives. Journal of Nervous and Mental Disease, 171(12), 742–748.

Sussman, R. W., & Cloninger, C. R. (2011). Origins of cooperation and Altruism. In R. H. Tuttle (Ed.), Developments in primatology: Progress and prospects. New York: Springer.

Walach, H., & Reich, K. H. (2005). Reconnecting science and spirituality: Toward overcoming a Taboo. Zygon: Journal of Religion & Science, 40(2), 423–439.

Preface

The understanding of mind and consciousness is one of the most exciting and challenging enterprises in the human's quest for comprehension of ourselves and of the universe as a whole. Chiefly, what is the nature of the mind and its relationship with the brain? What is it that makes us human and provides us with the qualities and skills that make us what we are? What is the source of the experience of ourselves? In spite of their importance, these questions remained largely neglected by philosophy and science during most part of the twentieth century. However, in the last 2 decades, there has been an exciting revival of interest in this subject in the academic milieu.

Discoveries in neuroscience and neurotechnology, in particular, have provided a unique window through which we can glance into the intricate workings of the human brain. Even though these technologies have evolved, they have also shown the fundamental limitations that currently exist in our understanding of the human mind. As put by the philosopher of mind David Chalmers (1995), despite the extraordinary advances of neuroscience, explaining conscious experience "poses the most baffling problems in the science of the mind" (p.200).

However, many people, even in the academic world, think that these questions have been already answered. They believe that the human brain is the answer, that mind does not exist, or it is just the product (for some, an epiphenomenon, an ineffective by-product) of brain chemistry and electric activity. Many also see the brain as an entity that can see, hear, think, feel, and make decisions. However, those seem to be unwarranted conclusions. As put by the neuroscientist Eccles (Popper and Eccles 1977:225):

> There is a general tendency to overplay the scientific knowledge of the brain, which, regretfully, also is done by many scientists and scientific writers. For example, we are told that the brain 'sees' lines, angles (…) and that therefore we will soon be able to explain how a whole picture is 'seen' (…). But this statement is misleading. All that is known to happen in the brain is that neurons of the visual cortex are caused to fire trains of impulse in response to some specific visual input.

A similar complaint was made by another couple composed of a philosopher and a neuroscientist who consider "the ascription of psychological – in particular, cognitive and cogitative – attributes to the brain is (…) a source of much (…) confusion.

(...) the great discoveries of neuroscience *do not require* this misconceived form of explanation" (Bennett and Hacker 2003:3–4).

Although reductionist materialism is a hypothesis worth pursuing, it is not a "scientific fact," as many believe. However, several reductionists accept that it is not yet a "scientifically proven fact," but it will become one soon. This belief that "at some unspecified time in the future" (p.205), it will be scientifically shown how brain generates mind is what Popper and Eccles called *promissory materialism*.

Of course that reductionism is a legitimate working theory regarding the mind-brain problem, however if it is hastily taken as the final and definitive answer, it might lead to a dogmatic and premature closure of this quest, which is one of the most important challenges to human knowledge. This approach is a dangerous epistemological posture, since the bare fact is that we are far from actually understand and explain mind. Using the terminology of the philosopher of science Thomas Kuhn (1970), we could say that we are in a preparadigmatic phase regarding the mind-brain problem. A preparadigmatic period is when there is no consensual acceptance by the scientific community of a specific paradigm (a framework of key theories, instruments, values and metaphysical assumptions for a given academic discipline) (Bird 2009). We have several candidates to be the scientific paradigm for the study of consciousness, but none have actually achieved that point yet, characterizing the field as an immature science.

One of the adverse consequences of the premature acceptance of a theory is that finding confirmatory examples of almost any theory is an easy task (Popper 1995). Much data is usually presented to support that mind has been fully explained as a product of brain activity. This often includes examples of psychophysiological concomitance and showing that brain injury or a neurophysiological change is often followed by some alteration in mind. However, as William James (1898) demonstrated more than a century ago, these data can also be accommodated by a *transmission theory* in which brain acts as a filter, having a "permissive or transmissive function" (p.291), acting as "an organ for limiting and determining to a certain form a consciousness elsewhere produced" (p.294). Also, as put by Chalmers (1995), studying neural correlates of consciousness, it is not the same as explaining consciousness or how and why these processes might give rise to conscious experience. There is an "*explanatory gap* between the functions and experience, and we need an explanatory bridge to cross it" (p.203).

According to the philosopher of science Karl Popper, to truly test a theory, we should be committed to look for evidence that could possibly falsify that theory. A good scientific theory withstands vigorous attempts to find contrary evidence. However, Kuhn (1970) showed that scientists usually are not able to recognize phenomena not allowed by the paradigm they are committed to:

> Can it conceivably be an accident, for example, that Western astronomers first saw change in the previously immutable heavens during the half-century after Copernicus' new paradigm was proposed? The Chinese, whose cosmological beliefs did not preclude celestial change, had recorded the appearance of many new stars in the heaven at a much earlier date (Kuhn 1970, p. 116).

The recognition that we are in a preparadigmatic phase in the exploration of the mind-brain problem would enable us to pursue a more fruitful investigation. It is worthwhile to remember that the scientific skills required to work in a preparadigmatic phase are different from those required during a paradigmatic phase, a period called by Kuhn as *normal science*. Fruitful work in preparadigmatic or revolutionary periods requires a more open-minded approach and not too strong commitment to any of the paradigm candidates. It would also require enlarging as much as possible the diversity of the empirical base and avoiding rushed rejection of hypotheses (Chibeni and Moreira-Almeida 2007).

A good scientific theory needs to be able to explain a wide and diversified range of phenomena (Hempel 1966). A theory based on a limited variety of phenomena has a very fragile base. The mere repetition of some sort of findings adds little strength and validity to a given theory. So, the deliberate search of new kinds of empirical observations to try a given paradigm is of great value because it may offer new and valuable confirmations, or, on the opposite, may lead to its rejection.

Throughout history, scientific revolutions often occurred when brilliant scientists took into account a wide range of previously unknown or dismissed phenomena. Galileo with his telescope and Charles Darwin during his 5 years long travel in the Beagle gathered an enormous mass of empirical evidence that were not available to most scientists at their time. The trip and the telescope allowed Darwin and Galileo to face a huge broadening of the empirical base, a base that could no longer be explained by the biological and astronomical established paradigms at their times. The end of those stories is well known to us. The same happened with classic physics, which, more than one century ago, seemed to be able to explain the whole nature. Such certainty made the eminent physicist Lord Kelvin state in 1900, a few years before Einstein developed relativity theory: "There is nothing new to be discovered in physics now, all that remains is more and more precise measurement." In fact, classic physics is very efficient in explaining most of the physical phenomena happening in our daily lives. However, when the study of microscopic particles and extreme velocities began, its limitation became evident, giving birth to the scientific revolution of modern physics (Greyson 2007).

So, the science and philosophy of mind need to enlarge their current timid scope and deal with a much wider range of phenomena if they in fact wish to make a truly significant contribution to the understanding of mind and its relationship with brain. In the exploration of the mind-brain problem, it is essential to take into consideration the whole range of human experiences, it does not matter how odd they may seem at first sight. Specifically, experiences often called "anomalous" and/or "spiritual" constitute a kind of empirical data that have been neglected in the last century, but with a high potential of being of enormous heuristic value (Cardeña et al. 2000, Eysenck and Sargent 1993, James 1909, Kelly et al. 2007). In order to not repeat the faults described above, we need to pay special attention exactly to the most extreme and challenging phenomena to advance our understanding. In this kind of exploration, it is necessary to give epistemological supremacy to consistent empirical data over any established or cherished theoretical hypothesis (Chibeni and Moreira-Almeida 2007), an approach in line with what was called by James (1976) as radical empiricism.

A whole range of human experiences that are at the core of spiritual traditions and beliefs have been neglected by academics, who refuse to take them seriously as empirical data that might shed light on the exploration of human nature. One possible explanation of this dismissal is the very common confusion between science and the metaphysical/philosophical positions of scientism and materialism. As John Haught (2005) discussed, although there is a widespread belief that science (a method of exploration) is inseparable from a materialistic ideology (a metaphysical proposition, a worldview), "it is not written anywhere that the rest of us who appreciate science have to believe that (materialist naturalism). In fact, most of the great founders of modern science did not. (...) [it] is not a scientific statement but a profession of faith" (p.367). Given the misguided conflation of science with materialism, it is understandable that most academic discussions avoid the investigation of experiences that might suggest a transcendental or nonmaterial reality, or, at least, take into consideration these phenomena as human experiences that deserve being studied in depth (Wallach and Reich 2005; Reich 2007). Actually, it is a mistake to take the materialistic worldview as a limitation or boundary for the scientific enterprise. Hefner (2006, 2007) and Helmut Reich (2007) convincingly argue for the enlargement of the empirical base for the scientific study of spiritual aspects of human experience, even (or mainly) if the observational data do not fit the existing mainstream (philosophical) framework.

It is also important not to reject an explanatory hypothesis because it is not fashionable or because it has been associated to superstition. Isaac Newton's formulation of gravity faced strong opposition because he was not able at that time (and we still are not able too) to explain how an object could influence another object at distance, with no material contact. This was even a more important problem since it was then prevalent two paradigms, mechanism and corpuscularianism, where the different properties of matter should be fully explained by the mechanical interactions of corpuscles (Blackburn 2008). Like Newton, Semmelweis and John Snow faced strong resistance and accusations of superstition and unscientific thinking by their contemporary scientists when proposing the contagion and germ theory, since these concepts were popular among superstitious and poorly educated people, while well educated people usually "knew" that miasma's theory was the truth (Lilienfeld 2000; Smith 2002; Vandenbroucke 2000).

Another kind of naïve epistemological prejudice is related to the rejection of qualitative data and the overemphasis on statistical analysis and quantitative data. It is often forgotten that one of the most important contemporary scientific paradigms, natural selection, emerged from qualitative studies performed by Charles Darwin (Ghiselin 1972). According the philosopher of science Alan Chalmers, people holding the idea that "if you cannot measure, your knowledge is meager and unsatisfactory" fail to "realize that the method that they endeavor to follow is not only necessarily barren and unfruitful but also is not the method to which the success of physics is to be attributed" (Chalmers 1978, p. xiv).

In the search for a paradigm to understand consciousness, it is necessary that it explains as much as possible the wide range of human experiences. It is essential to keep both intellectual humility and scientific rigor. As stated by Popper (1995), "in searching for the truth, it may be our best plan to start by criticizing our most cherished beliefs" (p.6).

Unfortunately, such open-minded approach is not always present in the history of science. Scientific revolutions did not triumph because the new paradigm was able to convert all skeptics and leaders of the opposition:

> The transfer of allegiance from paradigm to paradigm is a conversion experience that cannot be forced. Lifelong resistance, particularly from those whose productive careers have committed them to an older tradition of normal science, is … an index to the nature of scientific research itself. The source of resistance is the assurance that the older paradigm will ultimately solve all its problems….
>
> …[A] generation is sometimes required to effect the change…. Though some scientists, particularly the older and more experienced ones, may resist indefinitely, most of them can be reached in one way or another. Conversions will occur a few at a time until, after the last holdouts have died, the whole profession will again be practicing under a single, but now a different, paradigm. (Kuhn 1970, pp. 151–152)

The present book's main objective is to discuss the relationship between the mind and the brain from scientific and historical/philosophical perspectives. We focused on the discussion of topics about the mind-brain problem that are relevant, but usually neglected in academic debates. We have discussed basic concepts and empirical data that do not fit well in the reductionist hypothesis to explain the mind-brain problem. Most of chapters are further development of papers presented at the "Exploring the Frontiers of the Mind-Brain Relationship: An International Symposium" that was organized by this book's editors and took place in São Paulo (Brazil) in September 2010. This event, promoted by the Schools of Medicine of the Federal University of Juiz de Fora and of the University of São Paulo, put together several leading international researchers in the field, and proved to be an exciting and fruitful opportunity to rethink mind-brain relationship. The organizers and contributors of this work do not necessarily agree with all the positions expressed through the book, but are open enough to audaciously present and discuss arguments and data that too often do not have their deserved space in academic debate, which is supposed to be governed by freethinking and tolerance to divergences.

The book begins by addressing some theoretical (philosophical, historical, and physics) aspects related to common misconceptions, poorly known historical facts, and ingenious theories with the purpose of settling the debate in clearer and more solid grounds. This beginning shows that reductionist materialism is not the only rational and logic option and that other approaches are, at least, intellectually viable. It is followed by chapters presenting empirical data suggestive of nonreductionist views of mind. We believe that this book will provide an opportunity for a high-level debate of controversial and challenging topics in the quest for understanding the human mind.

The first two chapters discuss philosophical issues. They address several limitations of reductionist materialism and of arguments against nonmaterialist approaches. Saulo Araujo discusses in Chap. 1, historical and philosophical limitations of what he called "materialism's eternal return." He shows that the current and fashionable metaphors related to materialist reductionist explanations and the hope that in short period of time they would fully explain mental phenomena (promissory materialism) are an old phenomenon dating at least since the eighteenth century. Next, Robert Almeder, in Chap. 2, discusses and rejects five basic objections that materialists often raise to Cartesian Mind-Body dualism.

Carlos Alvarado, at Chap. 3, finishes the presentation of some theoretical background to enrich the historical and philosophical perspective in analysis the data that will be presented. He presents a historical overview of a productive and fruitful, but currently neglected, tradition of investigating and discussing the implications of psychic/anomalous phenomena to the mind-brain problem. His purpose is not to defend any specific position regarding the ontology of these experiences, but to show the relevance of them for our discussions.

The next couple of chapters, written by the physicist Chris J. S. Clarke (Chap. 4) and the physicians Stuart Hameroff and Deepak Chopra (Chap. 5), present models based on modern physics compatible with nonreductionist views of mind. Although quantum physics has been too often misused in discussions related to consciousness and spirituality by authors who actually have little acquaintance and expertise in this subject, this is certainly not the case with our three collaborators on this topic.

The next two chapters present and discuss data regarding neuroimaging studies, correcting several misunderstandings about the interpretations of this kind of findings to mind-brain problem. Jesse Edwards, Julio Peres, Daniel Monti, and Andrew Newberg (Chap. 6) competently review the increasing body of data related to neurophysiological studies on mindfulness. Mario Beauregard (Chap. 7) presents cutting-edge data regarding emotional regulation, suggesting that mind may act as an efficient cause, changing brain function. He also discusses the findings and implications of neurofunctional studies of spiritual experiences.

The last four chapters present and discuss the implications for the mind-brain problem of four types of human experiences that have often been called spiritual or anomalous exactly because they do not fit, *prima facie*, with reductionist materialist perspectives. However, despite how odd they may seem at first glance, they have been subject to in-depth studies by dozens of eminent scientists for more than a century. Two chapters deal with intriguing experiences happening near death and in the dying process. Peter Fenwick, a leading authority in studies of consciousness in dying people, wrote the chapter on near-death experiences (Chap. 8). He, with Franklin Santos, at the Chap. 9, discusses several kinds of end of life experiences and their theoretical and clinical implications.

Finally, Alexander Moreira-Almeida (Chap. 10) discusses studies about mediumship, an experience when a subject (the medium) claims to be in contact with deceased personalities, and Erlendur Haraldsson (Chap. 11) summarizes findings of investigations with children who claim to remember previous lives. Although most cases of both of these experiences in daily life do not present any challenging evidence, there are several well-documented cases that deserve further investigation.

This book is addressed to a wide academic audience, including philosophers, psychologists, physicians, neuroscientists, and all others interested in the mind-brain problem. But it may be also useful for nonacademic educated people interested in the subject. The book as a whole was conceived to be a rigorous and scientific one, but not hermetic; we have tried hard to make it accessible as much as possible to a broad range of readers. Given the specific interests of different readers, each chapter may be read separately – they can stand by themselves. However, as described previously, the book follows a logical sequence in which the previous chapters provide a foundation and background for a deeper analysis of what follows.

We hope you also may find this book useful and thought-provoking in the exploration of the frontiers of mind-brain relationship!

Juiz de Fora, Brazil   Alexander Moreira-Almeida, MD, PhD
São Paulo, Brazil   Franklin Santana Santos, MD, PhD

References

Bennett, M. R., & Hacker, P. M. S. (2003). Philosophical foundations of neuroscience. Malden: Blackwell.

BlackBurn, S. (2008). The Oxford dictionary of philosophy. Oxford: Oxford University Press.

Bird, A. (2009). *"Thomas Kuhn", The Stanford Encyclopedia of Philosophy (Fall 2009 Edition).* In N. Edward Zalta (Ed.). Available at http://plato.stanford.edu/entries/thomas-kuhn/#3

Cardeña, E., Lynn, S. J., & Krippner, S. (2000). *Varieties of anomalous experience: Examining the scientific evidence.* Washington: American Psychological Association.

Chalmers, A. F. (1978). *What is this thing called science? An assessment of the nature and status of science and its methods.* Indianapolis: Hackett.

Chalmers, D. J. (1995). Facing up to the problem of consciousness. *Journal of Consciousness Studies, 2*(3), 200–219.

Chibeni, S. S., & Moreira-Almeida, A. (2007). Remarks on the scientific exploration of "anomalous" psychiatric phenomena. *Revista de Psiquiatria Clínica, 34*(Suppl 1), 8–15.

Eysenck, H. J., & Sargent, C. (1993). *Explaining the unexplained: Mysteries of the paranormal.* London: England:Prion.

Ghiselin, M. T. (1972). *The triumph of Darwinian method.* Berkeley: University of California Press.

Greyson, B. (2007). Near death experiences: clinical implications. *Revista de Psiquiatria Clinica, 34*(suppl 1), 116–125.

Haught, J. F. (2005). Science and scientism: The importance of a distinction. *Zygon: Journal of Religion and Science, 40*, 363–368.

Hefner, P. (2006). Religion and science-two way traffic? *Zygon: Journal of Religion and Science, 41*, 3–6.

Hefner, P. (2007). Science and the big questions. *Zygon: Journal of Religion and Science, 42*, 265–268.

Hempel, C. G. (1996). *The philosophy of natural science.* Englewood Cliffs: Prentice-Hall.

James, W. (1898/1960). Human immortality: Two supposed objections to the doctrine. In G. Murphy & R. O. Ballou (Eds.), *William James on psychical research* (pp. 279–308). New York: Viking Press.

James, W. (1909/1960). The final impressions of a psychical researcher. In G. Murphy & R. O. Ballou (Eds.), *William James on psychical research* (pp. 309–325). New York: Viking Press.

James, W. (1976). *The works of William James – essays in radical empiricism.* Cambridge: Harvard University Press.

Kelly, E. F., Kelly, E. W., Crabtree, A., Gauld, A., Grosso, M., & Greyson, B. (2007). *Irreducible mind: Toward a psychology for the 21st century.* Lanham: Rowman & Littlefield Publishers.

Kuhn, T. S. (1970). *The structure of scientific revolutions.* Chicago: University of Chicago Press.

Lilienfeld, D. E. (2000). John Snow: The first hired gun? *American Journal of Epidemiology, 152*(1), 4–9.

Popper, K. R. (1995). Conjectures and refutations – the growth of scientific knowledge. London: Routledge.

Popper, K. R., & Eccles, J. (1977). *The self and its brain*. Berlin: Springer.

Reich, K. H. (2007). What needs to be done in order to bring the science-and-religion dialogue forward? *Zygon: Journal of Religion and Science, 42*, 269–272.

Smith, G. D. (2002). Commentary: Behind the Broad Street pump: Aetiology, epidemiology and prevention of cholera in mid-19th century Britain. *International Journal of Epidemiology, 31*(5), 920–932.

Vandenbroucke, J. P. (2000). Invited commentary: The testimony of Dr. Snow. *American Journal of Epidemiology, 152*(1), 10–12.

Walach, H., & Reich, K. H. (2005). Reconnecting science and religion: Toward overcoming a taboo. *Zygon: Journal of Religion and Science, 40*, 423–441.

Endorsements

Although dogmatic materialism is the monarch of contemporary Western philosophy and science, contributors to this splendid book remind me of the brash lad in the classic fable who shouted, "But the emperor has no clothes!" Some readers will agree with this observation while others will find it an outrageous heresy. But as they wend their way through each articulately stated and meticulously argued chapter, they will never succumb to boredom. It is the type of book that will haunt its readers long after the last chapter is read.

Stanley Krippner, PhD,
Professor of Psychology, Saybrook University,
Co-editor, *Debating Psychic Experience: Human Potential
or Human Delusion?*

This is really a thorough and up-to-date study of so-called anomalous spiritual phenomena. Instead of reducing these phenomena into exceptions to the general rule, the authors do not hesitate to widen the frame of interpretation. As is argued against a naïve epistemological stance, their open-minded inquiry is really a challenge and the authors managed to give us a fruitful perspective on their scholarly and multidisciplinary well documented efforts. I recommend this book to all psychiatrists, as well as professionals in theology and psychology of religion.

Peter J. Verhagen, M.D.,
Psychiatrist,
Chair of the World Psychiatric Association Section on Religion,
Spirituality and Psychiatry

It has become almost of a mantra in recent years to repeat that consciousness and mind are *caused* by brain processes and that it is only a matter of time before the neurosciences reveal all there is to know about them. As this important book shows, the fact that we now know a lot more about brain *correlates* of mental processes has not changed a number of awkward problems for a purely materialistic account of the mind, including how to solve the *hard problem* of consciousness and explain ordinary processes such as memory or less ordinary processes such as apparent psi phenomena. The contributors to this volume cover the essential areas in this discussion (philosophy, history, contemporary neuroscience, physics, and various anomalous experiences) and provide a cogent and scientifically based discussion that evidences that mind-brain relationships are far from definitely "explained" or obvious as a number of materialist authors have opined recently.

Etzel Cardeña, PhD,
Thorsen Professor, Lund University, Sweden,
Co-editor of *Altering Consciousness: A Multidisciplinary Perspective*
and *Varieties of Anomalous Experience.*

Acknowledgments

This book is the product of a collaboration of many people and organizations. Below we list most of them, but we are sure this list is incomplete.

We are deeply grateful to all the supporters of the "Exploring the Frontiers of the Mind-Brain Relationship: An International Symposium," the event where the idea of this book germinated. This event, promoted by the Federal University of Juiz de Fora (UFJF) and the University of São Paulo (USP), was organized by Pinus Longaeva Assessoria e Consultoria em Saúde e Educação Ltda, and received support from Hospital João Evangelista, Hospital Américo Bairral, Brazilian Association of Clinical Neuroscience (Abranec), Brazilian Association of Emergency Medicine (Abramed), Brazilian Association of Neuropsychology – Mid-west section, Editora Comenius, and Editora Atheneu.

At the Federal University of Juiz de Fora, we are grateful especially to Profs. Alexandre Zanini, Julio Chebli, and Klaus Alberto, great friends and supporters of this project. From the University of São Paulo, Prof. Wagner Gattaz has always given his support, and Prof. Homero Vallada has been a continuous source of ideas, stimulus, help, and friendship.

We also would like to thank the many professors and scholars who read in advance and made valuable suggestions to the book: the neurologists Mario Peres and Marcelo Maroco, the neurophysiologist Carlos Mourão Jr, the philosopher Rogério Severo, the philosophers and physicists Silvio Chibeni and Osvaldo Pessoa Jr., the physicists Alexandre Fonseca and Ademir Xavier, the historian and journalist Denise Paraná, and the scholar Alexandre Rocha.

Finally, we would like to thank Springer's editors Judy Jones and Garth Haller, who were very kind, helpful, and diligent throughout the entire book production process.

Contents

Contributors

Robert Almeder, PhD (USA)
Professor Emeritus of Philosophy at Georgia State University, and The McCullough Distinguished Visiting Professor, Hamilton College (2004–2006). Author of more than 80 articles published in academic journals and dozens of books. Former editor of the American Philosophical Quarterly from 1998 to 2003, published "Death and Personal Survival: the evidence for life after death" (Rowman and Littlefield, 1992). His most recent publication is "Truth and Skepticism" (Rowman and Littlefield, 2010).

Carlos S. Alvarado, PhD (USA)
Scholar in Residence and Faculty at Atlantic University, Assistant Professor of Research at the University of Virginia, and Adjunct Research Faculty at the Institute of Transpersonal Psychology. His research work has centered on survey research about out-of-body experiences and other psychic experiences, and studies about the history of psychical research. Alvarado is on the editorial boards of the Journal of Near-Death Studies and the Journal of the Society for Psychical Research, is an Associate Editor of the Journal of Scientific Exploration, and was the recipient of the 2010 Parapsychological Association's Outstanding Career Award. He is one of the editors of Research in Parapsychology 1993 (Lanham, MD: Scarecrow Press, 1998) and the author of over 200 articles published in psychology, psychiatry, and parapsychology journals.

Saulo de Freitas Araujo, PhD (Brazil)
Professor of History and Philosophy of Psychology at the Federal University of Juiz de Fora (Department of Psychology). Doctorate in philosophy (State University of Campinas/University of Leipzig). Research interest in history and philosophy of psychology, and the mind-brain problem in philosophy and psychology.

Mario Beauregard, PhD (Canada)
Associate Research Professor at the University of Montreal (Departments of Psychology and Radiology; Neuroscience Research Center; Centre de Recherche du Centre hospitalier de l'Université de Montréal). He is the author of more than 100 publications in neuroscience, psychology, and psychiatry. His main research interest is in neurobiology of emotion and mystical experience.

Deepak Chopra, MD (USA)

Author of more than 60 books on mind-body health, spirituality, and peace, 18 of which are New York Times bestsellers. Dr. Chopra's fiction and nonfiction books have been published in more than 85 languages. He is founder of the Chopra Foundation and founder and co-chairman of the Chopra Center for Wellbeing in Carlsbad, California. Dr. Chopra is a fellow of the American College of Physicians, a member of the American Association of Clinical Endocrinologists, an adjunct professor of executive programs at the Kellogg School of Management at Northwestern University, and a senior scientist with The Gallup Organization.

Chris Clarke (C.J.S. Clarke), PhD (UK)

School of Mathematics, University of Southampton, Southampton. Chris Clarke received his doctorate in mathematics at the University of Cambridge in 1970 and then worked in astrophysics and general relativity at Cambridge and Hamburg. After a period at the University of York he became Professor of Applied Mathematics at the University of Southampton in 1986, where he additionally worked on quantum cosmology and brain magnetometry, leaving in 1999 to work freelance with an honorary position at Southampton. He has published over 80 papers and book chapters, 6 single author books and 2 edited books in relativity, quantum theory, brain physics, philosophy, and religion.

Jesse Edwards (USA)

Medical student at Thomas Jefferson University Medical College and is working at the Myrna Brind Center of Integrative Medicine.

Peter Fenwick, MB BChir (cantab), DPM, FRCPsych (UK)

Department Neuroscience, Southampton University, UK. Honorary Senior Lecturer at the Kings College, Institute of Psychiatry, London University, UK. Dr Fenwick has a long standing interest in brain function and the problem of consciousness and has published a large number of research papers related to altered states of consciousness, and abnormalities of consciousness and behavior. One of his main interests for some years has been near-death experiences and the dying process.

Stuart Hameroff, MD (USA)

Professor of Anesthesiology and Psychology, and Director of the Center for Consciousness Studies at the University of Arizona in Tucson, Arizona. A clinical anesthesiologist, Hameroff also organizes the well-known interdisciplinary conferences "Toward a Science of Consciousness." He has collaborated with the eminent British physicist Sir Roger Penrose on the "Orch OR" theory of consciousness based on quantum computations in microtubules inside brain neurons. The theory suggests a connection between brain processes and fundamental spacetime geometry, the most basic level of the universe.

Erlendur Haraldsson, PhD (Iceland)

Professor emeritus, Department of Psychology, University of Iceland. Professor Haraldsson has published over 200 papers, on a broad range of issues, from standardization of psychological tests, to interrogative suggestibility, conducted national surveys of anomalistic and religious experiences, and done a major study of hallucinatory/visionary experiences in terminal patients. He has studied some one hundred cases of children who claim to remember episodes form a past life, and conducted psychological studies of them.

Daniel A. Monti, MD (USA)

Director of the Myrna Brind Center of Integrative Medicine at Thomas Jefferson University and Hospital in Philadelphia. He is also an Associate Professor in the Departments of Emergency Medicine and Psychiatry at Thomas Jefferson University. He is a leading scholar in the study of complementary and alternative medicine techniques and has recently published the book, The Great Life Makeover.

Alexander Moreira-Almeida, MD, PhD

Was trained in psychiatry and cognitive-behavioral therapy at the Institute of Psychiatry of the University of São Paulo, Brazil, where he also obtained his PhD in Health Sciences investigating the mental health of Spiritist mediums. Formerly a postdoctoral fellow in religion and health at Duke University, he is now Professor of Psychiatry at the Federal University of Juiz de Fora School of Medicine and Founder and Director of the Research Center in Spirituality and Health, Brazil (www.ufjf.br/nupes-eng). His main research interest involves empirical studies of spiritual experiences as well as the methodology and epistemology of this research field. His publications are available at www.hoje.org.br/elsh.

Andrew B. Newberg, MD (USA)

Director of Research at the Myrna Brind Center of Integrative Medicine at Thomas Jefferson University and Hospital in Philadelphia. He is also a Professor in the Departments of Emergency Medicine and Radiology at Thomas Jefferson University. He is a leading scholar in the neuroscientific study of different meditative practices and has recently published the book, Principles of Neurotheology.

Julio Peres, PhD (Brazil)

Clinical Psychologist, PhD in Neuroscience and Behavior, Institute of Psychology, University of São Paulo, Postdoctoral at the Center for Spirituality and the Mind, University of Pennsylvania, and at Radiology Clinic/Neuroimaging Diagnostic, Federal University of São Paulo (UNIFESP), Researcher at Institute of Psychiatry, University of São Paulo, Program of Health, Spirituality and Religiousness (PROSER).

Franklin Santana Santos, MD, PhD

Was trained in geriatrics at the Clinical Hospital of the University of São Paulo, Brazil, where he also obtained his PhD in Health Sciences investigating delirium in elderly patients. Formerly a postdoctoral fellow in cognitive disturbances at Karolinska Institute (Sweden), he is now Professor of postgraduate program of University of São Paulo School of Medicine and collaborator researcher of Laboratory of Neuroscience (LIM-27) at Institute of Psychiatry of the University of São Paulo, Brazil. He is a leader of the studies in issues related to death, dying, and palliative care in Brazil. His main research interest involves cognitive disturbances, thanatology, palliative care, and medical education, topics about which he has published several articles and books.

Part I
Philosophy and History

Chapter 1
Materialism's Eternal Return: Recurrent Patterns of Materialistic Explanations of Mental Phenomena

Saulo de Freitas Araujo

Abstract Since the new developments of neurotechnologies for studying the brain functioning in the second half of twentieth century, a new wave of enthusiasm for materialistic explanations of mental phenomena has invaded philosophy and psychology departments worldwide. The culmination of all this was the so-called "decade of the brain" in the 1990s. However, a closer examination of the arguments presented by some of these new materialists reveals recurrent patterns of analogies and metaphors, besides an old rhetorical strategy of appealing to a distant future, in which all the problems will be solved. This study intends to show that these new forms of materialism repeat discursive strategies of older versions of materialism, especially the French materialism of the eighteenth century and the German materialism of the nineteenth century. Finally, an interpretation for materialism's eternal return will be offered.

> For what can be more harmful to knowledge than falsely communicating even mere thoughts, than concealing doubts which we feel about our own assertions, or giving a semblance of self-evidence to grounds of proof which do not satisfy ourselves? (Immanuel Kant)
>
> When they say that matter is the substance and cause of all the phenomena, but do not give a satisfactorily clear concept either of matter or of the manner in which everything arises from it, then their materialism is little more than unintelligible talk, as dark and incomprehensible as the suprasensual assumptions of their opponents. (Heinrich Czolbe)

For a significant part of our modern society, materialism seems to be the natural and inevitable result of the advancement of scientific research. In fact, this seemingly flawless and often propagated image of an identity between science and the materialist worldview is not without some legitimacy, to the extent that a significant number of scientists make a point of stating their belief in materialism, and devote their time to the popularization of this idea. However, this image does not exactly correspond to reality. Examined more closely, it proves to be very limited and

S. de Freitas Araujo (✉)
Departamento de Psicologia, Universidade Federal de Juiz de Fora,
Caixa Postal 611, Juiz de Fora, MG 36001-970, Brazil
e-mail: saulo.araujo@ufjf.edu.br

A. Moreira-Almeida and F.S. Santos (eds.), *Exploring Frontiers of the Mind-Brain Relationship*, Mindfulness in Behavioral Health, DOI 10.1007/978-1-4614-0647-1_1,
© Springer Science+Business Media, LLC 2012

problematic, since there are also many scientists who clearly speak out against the materialist worldview, thereby demonstrating the independence of science and materialism.

My aim in this study is to debunk this false identity and to show how it has generated mythical views on human nature, taking as examples the more radical attempts in contemporary neuroscience to eliminate the autonomy of subjective human experience. This usually happens in three steps: initially, some capacities, previously assigned to human beings, are attributed to the brain or part of it; then, a complete physicalization of human nature in general is proclaimed, which is thus reduced to a mere product of the brain activity; and finally, this materialistic view is propagated as the inevitable result of contemporary science.

To achieve the aforementioned goal, this article is divided into four sections. Section 1.1 establishes some conceptual definitions to facilitate both the understanding of our central idea and its subsequent discussion. Section 1.2 shows how a significant part of contemporary neuroscience is committed to materialism and to the promise of a new future for mankind. Section 1.3 shows the great similarity between some arguments of contemporary materialists and the metaphors employed by materialists of the eighteenth and nineteenth centuries, thus characterizing what I call the "eternal return of materialism." Finally, Sect. 1.4 argues that the uncritical, naïve assimilation and reproduction of contemporary science favors the creation and propagation of myths and ideologies, against which we must be ever vigilant.

1.1   Conceptual Definitions

Before setting out the central idea, I will begin with some conceptual clarifications to facilitate both understanding and discussion.

Materialism, though it now appears closely associated with the contemporary image of science, is not itself a scientific theory. In its most general sense, it is a metaphysical thesis about the ultimate nature of reality, which unifies the whole field of human experience, reducing it ultimately to some explanatory principle derived (valid or invalidly) from the concept of matter, and finally providing a worldview. In other words, the hallmark of each and every advocate of materialism is his or her commitment to the thesis that everything that exists in the world is material.[1] However, such a statement is so general and comprehensive that it can never be subjected to any particular empirical test, thus going far beyond the sphere of any possible scientific knowledge. Indeed, at least since Kant's *Critique of Pure Reason* (1781/ 1998), it is no longer possible to ignore that the totality of the world is just a rational idea, and not an object that can be given in our experience. The same applies to matter, which, if thought of as the ultimate condition of experience,

---

[1] One can argue, of course, that there are weaker forms of materialism, according to which there are also numbers and other abstract entities that cannot be reduced to material processes or states. However, in what sense they should be properly called "forms of materialism" is not at all clear to me. In any case, my arguments are directed only to materialism in its stronger forms.

cannot be confused with any particular empirical phenomenon, since in this case it would have to be explained by something other than itself. That is to say, if matter is thought of as the absolute foundation of all human experience, it can never appear as an object of our experience, thus remaining beyond the reach of our scientific knowledge. Therefore, since science must be supported by empirical evidence, no scientific theory, however well established, may imply materialism. And that is exactly why we cannot confuse it with a scientific theory and treat it as if it were one. Therefore, the expression "scientific materialism" can serve at most to designate the ideological stand or the professional status of those who believe in it (scientists), but in no way does this entail that they are proposing a scientific theory. This confusion being unraveled, it should now be clear that the legitimacy of science does not depend on scientists' commitment to materialism, but only on their commitment to logic and scientific methodology. Thus, it should be noted that all that science can do is to discover the existence of phenomena and their relations, but never the essence and ultimate nature of reality, since this cannot be given in the empirical level and would thus require a different kind of knowledge. Moreover, although scientific practice can be attached to a worldview, as Fleck (1979) and Kuhn (1970) have showed, science is an epistemic activity in constant development, so that its crystallization in a worldview would be contrary to its own nature. In short, materialism and science are different things, which only due to a conceptual mistake can be treated as identical.

This first characterization of materialism is still insufficient to support the central idea of this paper. So it is necessary to introduce a second clarification, which concerns the diversity of ways in which it has appeared in the Western intellectual tradition. In fact, there are specific contextual differences related to the emergence of each type of materialism. But there are also important similarities between its various manifestations. This is what sustains the classic statement of Lange (1866), according to which materialism is as old as philosophy itself. However, as there are many different conceptions of what constitutes the ultimate reality of matter and also of its main explanatory principle, materialism finds different forms of expression over time – not to mention the materialists who do not even present a definition of matter, thus making the term even more vague and comprehensive. Its first manifestation occurs among the ancient Greeks, with the atomism of Leucippus and Democritus, later revived by Epicurus, whose goal was to explain all the reality in terms of size, shape, and motion of atoms in a vacuum. But from the atoms of Democritus and Epicurus to contemporary physics, the term "matter" has taken on a plurality of meanings, which consequently must lead to different kinds of materialism.[2] We should, therefore, always be alert to the different meanings that the terms "matter" and "materialism" may take in contemporary debates – and especially in so far as there are forms of materialism which, in the absence of an explicit concept of matter, rely on a particular physical entity (e.g., brain) or on an explanatory principle (e.g., natural selection). This whole range of potential uses of a single word allows

---

[2] The conceptual changes in physics over time offer a good indication of the complexity and the problems involving the notion of matter (Jammer 1961, 1966, 2000).

us to glimpse the broad semantic spectrum of materialism (Bloch 1972; Campbell 1967; Lange 1866; Moser and Trout 1995).

After these initial considerations, I would like to define the type of materialism that I will take as my object of analysis. Leaving aside its more archaic forms, I will focus here on modern materialism, which arose during the second half of the eighteenth century linked to the progress of natural science and remains alive and well in the present (Nieke 1980; Bayertz 2007). Within this context, what specifically interests me here is a kind of materialism associated with contemporary neuroscience, which takes the mind to be nothing but the brain, thus elevating the brain to an omnipotent physical entity and seeking in its properties the fundamental explanatory principle of mental phenomena. In other words, my emphasis will be on materialism not as a general theory of reality, not as a worldview, but only as a strategy for explaining our subjective experience.

1.2   Materialism and Contemporary Neuroscience

In the last decades we have witnessed a renewed interest in materialism. A good measure of this fact is the increasing number of recent scholarly books on the subject, which have sought to show both its origins and its relevance to the present (Audidière et al. 2006; Arndt and Jaeschke 2000; Bayertz et al. 2007; Boulad-Ayoub and Torero-Ibad 2009; Moser and Trout 1995). Moreover, in the very field of philosophy of mind, a discipline that deals more specifically with the mind-brain problem, there are several recent attempts to explain the human mind in materialistic terms (e.g., Melnyk 2003; Papineau 2002; Ramachandran 2004; Tye 2009). Not to mention the recent "manifestos" in defense of materialism, which play a broader cultural role, by appealing to a complete conversion of the reader to atheism (Dawkins 2006; Dennett 2006; Harris 2006; Hitchens 2007; Onfray 2005). This renders, in my view, the thesis of a weakening of materialism (Koons and Beeler 2010) highly problematic.

More specifically, we have witnessed a wave of enthusiasm motivated by the continuous and remarkable advances in neuroscience, particularly the new neuroimaging technologies. Inside and outside the universities, new centers of study and/or research of brain processes arise. An explicit optimism underlies this wagering on brain research, seeing in a not too distant future the solution of various problems concerning human nature. We should not lose sight of the enthusiasm with which the international scientific community announced the beginning of "the decade of the brain" (1990–1999) and its promises to the fields of philosophy, psychology, and psychiatry.

Looking more closely at that general enthusiasm, one can see that one of its key features is the fact that its representatives are convinced of living in the dawn of a real revolution in human thought, which will radically transform the vision we have of ourselves. This becomes clear in the statements of some of the most prestigious representatives of contemporary neuroscience and philosophy:

> The advent of novel experimental approaches and imaging techniques is sure to transform our understanding of the human brain. What a unique privilege it will be for our

generation—and our children's—to witness what I believe will be the greatest revolution in the history of the human race: understanding ourselves. (Ramachandran and Blakeslee 1998, p. xvii)

Will a proper theory of brain function present a significantly different or incompatible portrait of human nature? Should we prepare ourselves, emotionally, for yet another conceptual revolution, one that will touch us more closely than ever before? [...] I am inclined toward positive answers to all of these questions, and toward an optimistic estimate of our future prospects, both scientific and moral. (Churchland 1996, p. 18-19)

As we investigate, however, the foundations of all this optimism, we find that it relies primarily on a general dissatisfaction with all forms of traditional psychological language (mind, belief, desire, will, intention, etc.) that these authors ascribe to the ingenuity of our Western religious and dualistic tradition. They argue that common sense created illusions that were subsequently corrected by scientific advances (e.g., the case of witches), and the same will likely happen with the advances of neuroscience, which will prove to be a superior theory of human nature. The following passage illustrates this idea:

Psychosis is a fairly common affliction among humans, and in earlier centuries its victims were standardly seen as cases of demonic possession, as instances of Satan's spirit itself [...] That witches exist was not a matter of any controversy. [...] But observable or not, we eventually decided that witches simply do not exist. We concluded that the concept of a witch is an element in a conceptual framework that misrepresents so badly the phenomena to which it was standardly applied that literal application of the notion should be permanently withdrawn. Modern theories of mental dysfunction led to the elimination of witches from our serious ontology. The concepts of folk psychology – belief, desire, fear, sensation, pain, joy, and so on – await a similar fate, according to the view at issue. And when neuroscience has matured to the point where the poverty of our current conceptions is apparent to everyone, and the superiority of the new framework is established, we shall be able to set about reconceiving our internal states and activities, within a truly adequate conceptual framework at last. Our explanations of one another's behavior will appeal to such things as our neuropharmacological states, the neural activity in specialized anatomical areas, and whatever other states are deemed relevant by the new theory. Our private introspection will also be transformed, and may be profoundly enhanced by reason of the more accurate and penetrating framework it will have to work with – just as the astronomer's perception of the night sky is much enhanced by the detailed knowledge of modern astronomical theory that he or she possesses. (Churchland 1988, p. 44-45)

One of the central problems of this kind of materialism is that in the absence of consistent empirical data, its supporters are forced to appeal to a very distant, almost mythical future, in which all their hopes will become reality. Moreover, this appeal to the future is never followed up by a more precise specification of the necessary and/or sufficient conditions for the realization of the announced revolution, or even by spelling out a concrete situation that could falsify its proposal. This raises the suspicion that we are dealing here with mere rhetoric, but in no way with an actual scientific hypothesis.[3] Thus, all that is left for us to believe is a promise with no

---

[3] The use of rhetorical strategies in the legitimation of scientific theories seems to be an essential feature of science, as recent studies have shown (Fahnestock 2002; Gross 1996, 2006; Irion 2008; Moss 1993; Shea 2009; Stark 2009). However, this does not entail that science is mere rhetoric or that one can do science with rhetoric alone.

deadline or expiration date. Metaphorically speaking, it is as if we had to buy a promissory note without knowing when we could redeem it. In other words, no more than a prophecy is offered, based only on the hope, or faith, that a revolution will happen in the future. A few decades ago, Popper had already noted the fundamental problem with this materialistic strategy:

> Promissory materialism is a peculiar theory. It consists, essentially, of a historical (or historicist) prophecy about the future results of brain research and their impact. This prophecy is baseless. No attempt is made to base it upon a survey of recent brain research. […] No attempt is made to resolve the difficulties of materialism by argument. No alternatives to materialism are even considered. (Popper 1977, p. 97)

The lack of a solid basis for the materialistic proposal becomes more evident when we analyze its relationship with the advancement of empirical investigations. In fact, after a few decades since the beginning of this general enthusiasm, we can ask if there is by now at least an outline of what should be the new neuroscientific theory of human nature. By way of illustration, in a recent anthology of interviews with some of the foremost scholars of the field – including the leading exponents of contemporary materialism (e.g., Crick, Dennett, and the Churchlands) – we find that after 40 years of neurophysiological research, we still do not have the least idea of how to solve the most basic problems about human consciousness. However, we are not left without guesses and/or general and vague opinions, in most cases incompatible with each other (Blackmore 2006). And when these authors are asked if there is still a long time until the promise can be realized, the most accurate response we receive comes from Patricia Churchland herself: "We don't really know how long we'll have to wait" (Churchland and Churchland 2006, p. 56).

Let me explore this point further. While the dreamed future supertheory does not come, what the materialists have to offer is a metaphorical, almost fictional language, through which they attach to the brain or parts of it a series of skills and accomplishments that were previously attributes of human beings as a whole. Thus, the "astonishing hypothesis," according to which we are just a bundle of neurons or the sum of the behavior of neurons and their molecules, is produced (Crick 1995). And ideas such as "the emotional brain" (Le Doux 1998), "the volitional brain" (Libet et al. 1999), "the executive brain" (Goldberg 2001), "the believing brain" (Gazzaniga 2005), "the brain that changes itself" (Doidge 2009), etc., become possible, too. Let us see some examples of how these metaphors are employed:

> My first assumption was that part of one's brain is concerned with making plans for future actions, without necessarily carrying them out. […] My second assumption was that one is not conscious of the 'computations' done by this part of the brain but only of the 'decisions' it makes – that is, its plans. […] Then, such a machine will appear to itself to have Free Will, provided it can personify its behavior – that is, it has an image of 'itself'. (Crick 1995, p. 266)
>
> The left hemisphere, in other words, was making emotional judgments without knowing what was being judged. The left hemisphere knew the emotional outcome, but it did not have access to the processes that led up to that outcome. As far as the left hemisphere was concerned, the emotional processing had taken place outside its realm of awareness (which is to say, had taken place unconsciously). (Le Doux 1998, p. 15)
>
> Let me be as clear as I can about what I mean by 'holding beliefs' or having belief systems. […] Overall, and this is my view about the nature of beliefs, our species instinctively

reacts to events, and in a specialized system of the human brain that reaction is interpreted. Out of that interpretation, beliefs emerge about rules to live by. [...] We now know that the left hemisphere of the brain – the one that attaches a story to input from the world – creates these beliefs. [...] It follows from the idea that if the brain is modular, a part of the brain must be monitoring all the networks' behaviors and trying to interpret their individual actions in order to create a unified idea of the self. Our best candidate for this brain área is the 'left hemisphere interpreter'. [...] I have called this area of the left hemisphere the interpreter because it seeks explanation for internal and external events and expands on the actual facts we experience to make sense of, or *interpret*, the events o four life. (Gazzaniga 2005, p. 146–148)

We then discover, not without astonishment, that the pinnacle of our scientific progress consists in replacing the notion of subject with the cerebral hemisphere. Who now "knows," "makes assessments," "interprets," "creates," "seeks explanation," etc. is no longer a person but a piece of matter (part of the brain). It turns out that the use of such enchanted metaphors, taken as real explanations of phenomena, produces the opposite result of what was promised, namely, a scientific explanation. After all, we must not forget that analogies and metaphors are valid only when, as well as the manifest similarities, the differences are also highlighted. If the latter disappear, then the relationship becomes one of identity and no longer of analogy.[4] As I tried to show elsewhere (Araujo 2003, 2006), the attribution of real properties and psychological skills to physical objects is characteristic of a very primitive stage of human intelligence, namely, the animism that every materialist wants to combat. And what is worse, this represents a return to a much less critical and more naïve metaphysics than the one they intend to overcome. We are dealing here, therefore, with pseudo-explanations, which in no way address the fundamental issues, and whose function is, again, merely rhetorical.

1.3   The Eternal Return

If we add to the analysis so far conducted an historical perspective, this new enthusiasm for the advancement of neuroscience reveals an even more interesting facet, namely, its similarity to certain aspects of earlier versions of materialism, which puts under suspicion the question of its novelty. As stated earlier, the association of materialism with the natural sciences, specifically with physiology, is a phenomenon that arises in the eighteenth century, especially in the context of the French Enlightenment. Later, in the German tradition, by the mid-nineteenth century, it will lead to the so-called "materialism dispute" (*der Materialismusstreit*) (Arndt and Jaeschke 2000; Bayertz et al. 2007; Gregory 1977; Meschede 1980; Wittkau-Horgby 1998).

---

[4] One cannot doubt that science has developed with the help of metaphors and analogies (Baake 2003; Brown 2008; Hallyn 2000; Hesse 1966; Leary 1990; Leatherdale 1974), but scientific theories do need far more than figurative language to be established.

Let us consider, first, how the brain is already present in these contexts as the central element in the explanation of mental phenomena:

> Since all the faculties of the soul depend so much from the specific organization of the brain and of the whole body, being nothing but this organization itself, we are dealing here with a well enlightened machine! (La Mettrie 1748, p. 70)[5]
>
> All the intellectual faculties, that is to say, all the modes of action attributed to the soul, may be reduced to the modifications, to the qualities, to the modes of existence, to the changes produced by the motion of the brain, which is visibly in man the seat of feeling – the principle of all his actions. [...]This brain moves itself in its turn, reacts upon itself. (Holbach 1770, p. 63)
>
> The operations of the soul or spirit result from movements executed by the cerebral organ. (Cabanis 1805, p. 40)
>
> One cannot doubt that the seat of consciousness, of will, of thought must be finally sought only in the brain. Only that, for now, we have been unable to determine the manner in which the machine gears are interrelated. (Vogt 1847, p. 17)[6]
>
> That the brain is the organ of thought, and that both are in an immediate and necessary relation, that one does not exist and cannot be thought without the other, is a truth which a physician or physiologist can hardly doubt. (Büchner 1855, p. 423)

Even more interesting is the profusion of metaphors and analogies created to explain how the brain can produce what we call mind. There appears here the traditional idea that mental processes are functions or products of brain activity, which is very much alive in contemporary debate[7]:

> In the same way as a violin string or a harpsichord key vibrates and gives out a sound, so the strings of the brain, struck by rays of sound, are stimulated to give out or repeat the words which touch them. (La Mettrie 1748, p. 34)
>
> To form an accurate idea of the operations from which thought results, it is necessary to consider the brain as a special organ designed especially to produce it, as the stomach and the intestines are designed to make the digestion, the liver to filter bile, the parotids and maxillary and sublingual glands to prepare the salivary juices. (Cabanis 1805, p. 152-153)
>
> I think that every natural scientist, who thinks in a logical way and with consistency, will come to the conclusion that all those capacities that we apprehend under the concept of mental activities are only functions of the brain substance; or, to express myself here in a more rudimentary way, that thoughts relate to the brain in the same way as the bile to the liver or urine to the kidneys. (Vogt 1847, p. 17)
>
> Thought is a motion of matter. (Moleschott 1852, p. 284)
>
> Just as the steam engine produces motion, the organic complex of evolved matter with energy potential generates in the animal body a sum of effects that, united, we call spirit, soul, thought. (Büchner 1855, p. 443)

These passages, taken here as an illustration only, are enough to show that contemporary materialists, by announcing their ideas as a great novelty, repeat discursive strategies and forms of reasoning very similar to those of their coreligionists in

---

[5] For the translations of the French citations, I adopted two different procedures: in the case of Holbach, I used a corresponding English edition of the original work. As for La Mettrie and Cabanis, the translations are all mine.

[6] The translations of the German citations are all mine.

[7] John Searle, for example, in one of his most famous books, has stated: "In my view we have to abandon dualism and start with the assumption that consciousness is an ordinary biological phenomenon comparable with growth, digestion, or the secretion of bile." (Searle 1997, p. 6).

the past, without being aware thereof. And even though there is no uniformity or consensus between them on what precisely is the brain, as shown by the recent history of neuroscience (e.g., Hagner 2000), it is the general attitude to it – as well as the weakness of the arguments and lack of empirical evidence supporting it – that has been repeated, and that I want here to emphasize. It is as if the entire period of our intellectual history that goes from the second half of the eighteenth century until the late nineteenth century had not existed, so that this new materialistic dawn may sound like something really new. In fact, however, this is just new clothing to old ideas and hopes, which at the end turned into an article of faith. It is this cyclical phenomenon, which appears in our culture over the past three centuries, that I am calling the "eternal return of materialism".

It is time, however, to ask: is there a meaning in this eternal return? At first, we can understand it only as a theoretical-conceptual naivety, which is due to ignorance or contempt in relation to the history of science and of philosophy. But my thesis is that it reveals something deeper, which concerns a lack of attention to the epistemic limits of human beings. In other words, what materialists are trying to do, at least since the eighteenth century, is to eliminate the autonomy of the subjective dimension of human experience, reducing it or reformulating it in terms of the objective sphere of natural sciences. In the language of contemporary philosophy of mind, this means explaining first person experiences from a third person perspective, an attempt that has so far been proved unsuccessful (Chalmers 1996, 2006; Frank 1995; Henrich 2007; Jackson 1982; Nagel 1974, 1986). Now, the abstraction of the subject in natural sciences is only a methodological tool, useful for the development of robust physical theories, but by no means a proof of its nonexistence or of its irrelevance to the understanding of human nature. It is as if two zoologists, swimming in a lake infested with hungry crocodiles, believed that, by stopping talking of crocodiles, they would be eliminating the imminent danger of being devoured by them. It is precisely this effort to eliminate the realm of subjective autonomy that has failed over time. In this sense, the eternal return of materialism reveals an eternal oblivion: materialists (scientists and philosophers) have forgotten the limits of human knowledge, not just merely of its empirical content, but of its universal conditions. In fact, the subjective experience cannot be eliminated or translated into a purely objective language, although it may be correlated to the latter. In this case, it would be useless to accumulate new empirical data to resolve the impasses and paradoxes of contemporary materialism (Araujo 2003). This allows us to suspect that the new vision of human nature that these materialists want to achieve is perhaps utopian and illusory, since the limits of the conditions for scientific knowledge seem to remain unchanged.

1.4 Concluding Remarks

I would like now to explore briefly the question of the relationship between contemporary science and the creation of ideologies. First, it is important to reaffirm that materialism is not a logical and inevitable consequence of scientific research; otherwise, there could be no antimaterialist scientists. In fact, however, many scientists since the nineteenth century have pointed out the impasses for materialism and the impossibility of solving them (e.g., Du Bois-Reymond 1872; Wundt 1889), and also many prestigious contemporary neuroscientists are antimaterialists (e.g., Beauregard and O'Leary 2007; Eccles 1977, 1980; Penfield 1975; Wallace 2000). Second, the naive and thoughtless exaltation and/or reproduction of a false ideal of science eventually leads to the creation of myths that hamper the understanding of what the scientific activity really is and how it developed historically, and cause scientists themselves to behave unscientifically (Midgley 1994, 2003; Numbers 2009). And it is in this context of an uncritical assimilation of contemporary science that we run the risk of accepting an ideology (scientism, scientific materialism) as if it were a genuine scientific product, and of participating in an ideological crusade without knowing that it is one. Now, if science has a primary function in our society, it consists precisely in the promotion of a critical examination and understanding of reality, not in the creation of fantastic tales and alienating myths. And if we cannot find definitive answers to certain questions that have been consistently raised up over time, this might point to certain limits of our knowledge, which forces us to constantly remind ourselves of the remaining obstacles to avoid the risk of falling into new forms of dogmatism. So, if the dualist metaphysics is only a vestige of our theoretical naivety and of our epistemic ignorance, why should we judge as less naive a doctrine that attributes magical properties to a physical object (the brain) and pin all the future hopes on it, thus creating a new fetish? In contemporary neuroscientific discourse, there seems to be actually less science than one usually imagines.

Acknowledgments I would like to thank Alexander Moreira-Almeida, Chris Clarke, Luis Henrique Dreher and an anonymous reviewer for helpful comments on an earlier version of the manuscript, as well as FAPEMIG (Research Foundation of the State of Minas Gerais) for financial support.

References

Araujo, S. F. (2003). *Psicologia e neurociência: uma avaliação da perspectiva materialista no estudo dos fenômenos mentais*. Juiz de Fora: Editora UFJF.

Araujo, S. F. (2006). Wie aktuell ist Wilhelm Wundts Stellung zum Leib-Seele-Problem? *Schriftenreihe der Deutschen Gesellschaft für Geschichte der Nervenheilkunde, 12*, 199–208.

Arndt, A., & Jaeschke, W. (Eds.). (2000). *Materislismus und Spiritualismus. Philosophie und Wissenschaften nach 1848*. Hamburg: Meiner.

Audidière, S., Bourdin, J.-C., Lardic, J.-M., Markowits, F., & Zarka, Y. C. (Eds.). (2006). *Matérialistes français du XVIIIe siècle. La Mettrie, Helvétius, d'Holbach*. Paris: Presses Universitaires de France.

Baake, K. (2003). *Metaphor and knowledge*. Albany: State University of New York Press.

Bayertz, K. (2007). Was ist moderner Materialismus? In K. Bayertz, M. Gerhard, & W. Jaeschke (Eds.), *Der Materialismus-Streit* (pp. 50–70). Hamburg: Meiner.

Bayertz, K., Gerhard, M., & Jaeschke, W. (Eds.). (2007). *Der Materialismus-Streit*. Hamburg: Meiner.

Beauregard, M., & O'Leary, D. (2007). *The spiritual brain. A neuroscientist's case for the existence of the soul*. New York: HarperCollins.

Blackmore, S. (Ed.). (2006). *Conversations on consciousness*. Oxford: Oxford University Press.
Bloch, E. (1972). Das Materialismusproblem. Seine Geschichte und Substanz. Frankfurt a.M.: Suhrkamp.
Boulad-Ayoub, J., & Torero-Ibad, A. (Eds.). (2009). *Matérialismes des modernes. Nature et moeurs*. Quèbec: Les Presses de l'Université Laval.
Brown, T. (2008). *Making truth: Metaphor in science*. Urbana: University of Illinois Press.
Büchner, L. (1855/1971). Kraft und Stoff. In D. Wittich (Ed.), *Schriften zum kleinbürgerlichen Materialismus in Deutschland, Bd 2* (pp. 343–516). Berlin: Akademie.
Cabanis, P. (1805). Rapports du physique et du moral de l'homme. Seconde Édition. Tome Premier. Paris: Crapelet.
Campbell, K. (1967). Materialism. In P. Edwards (Ed.), *The encyclopedia of philosophy* (Vol. 5, pp. 179–188). New York: Macmillan Publishing.
Chalmers, D. (1996). *The conscious mind. In search of a fundamental theory*. Oxford: Oxford University Press.
Chalmers, D. (2006). David Chalmers. In S. Blackmore (Ed.), *Conversations on consciousness* (pp. 37–49). Oxford: Oxford University Press.
Churchland, P. (1988). *Matter and consciousness. Revised edition*. Cambridge: The MIT Press.
Churchland, P. (1996). *The engine of reason, the seat of the soul. A philosophical journey into the brain*. Cambridge: The MIT Press.
Churchland, P., & Churchland, P. (2006). Patricia & Paul Churchland. In S. Blackmore (Ed.), *Conversations on consciousness* (pp. 50–67). Oxford: Oxford University Press.
Crick, F. (1995). *The astonishing hypothesis. The scientific search for the soul*. New York: Touchstone.
Dawkins, R. (2006). *The God delusion*. London: Bantam Press.
Dennett, D. (2006). *Breaking the spell. Religion as a natural phenomenon*. London: Penguin.
Doidge, N. (2009). *The brain that changes itself*. London: Penguin.
Du Bois-Reymond, E. (1872/1974). Über die Grenzen des Naturerkennens. In *Vorträge über Philosophie und Gesellschaft* (pp. 54–77). Berlin: Akademie.
Eccles, J. (1977/1983). Part II. In K. Popper, & J. Eccles (Eds.), *The self and its brain. An argument for interactionism* (pp. 225–422). New York: Routledge.
Eccles, J. (1980). *The human psyche*. Berlin: Springer.
Fahnestock, J. (2002). *Rhetorical figures in science*. Oxford: Oxford University Press.
Fleck, L. (1979). *Genesis and development of a scientific fact*. Chicago: The University of Chicago Press.
Frank, M. (1995). Is subjectivity a non-thing, an absurdity? On some difficulties in naturalistic reductions of self-consciousness. In K. Ameriks & D. Sturma (Eds.), *The modern subject* (pp. 177–197). Albany: State University of New York Press.
Gazzaniga, M. (2005). *The ethical brain*. New York: Dana Press.
Goldberg, E. (2001). *The executive brain. Frontal lobes and the civilized mind*. Oxford: Oxford University Press.
Gregory, F. (1977). *Scientific materialism in nineteenth century Germany*. Dordrecht: Reidel.
Gross, A. (1996). *The rhetoric of science*. Cambridge: Harvard University Press.
Gross, A. (2006). *Starring the text. The place of rhetoric in science studies*. Carbondale: Southern Illinois University Press.
Hagner, M. (2000). Homo cerebralis. Der Wandel vom Seelenorgan zum Gehirn. Frankfurt a.M.: Insel.
Hallyn, F. (Ed.). (2000). *Metaphor and analogy in the sciences*. Dordrecht: Kluwer.
Harris, S. (2006). *Letter to a Christian nation*. New York: Alfred Knopf.
Henrich, D. (2007). *Denken und Selbstsein. Vorlesungen über Subjektivität*. Frankfurt a.M: Suhrkamp.
Hesse, M. (1966). *Models and analogies in science*. Notre Dame: University of Notre Dame Press.
Hitchens, C. (2007). *God is not great. How religion poisons everything*. New York: Twelve.
Holbach, B. (1770/2001). *System of nature or laws of the moral and physical world*. Kitchener: Batoche Books.

Irion, C. (2008). *Charles Darwin's "The Origin of Species": Science, rhetoric and revolution.* München: Grin.

Jackson, F. (1982). Epiphenomenal qualia. *Philosophical Quarterly, 32,* 127–136.

Jammer, M. (1961). *Concepts of mass in classic and modern physics.* Cambridge: Harvard University Press.

Jammer, M. (1966). *The conceptual development of quantum mechanics.* New York: McGraw-Hill.

Jammer, M. (2000). *Concepts of mass in contemporary physics.* Princeton: Princeton University Press.

Kant, I. (1781/1998). Kritik der reinen Vernunft. In J. Timmermann (Ed.), Hamburg: Meiner.

Koons, R., & Beeler, G. (Eds.). (2010). *The waning of materialism.* Oxford: Oxford University Press.

Kuhn, T. (1970). *The structure of scientific revolutions* (2nd ed.). Chicago: The University of Chicago Press.

La Mettrie, J. O. (1748). L'Homme machine. Leiden: Elie Luzac.

Lange, F. (1866). *Geschichte des Materialismus und Kritik seiner Bedeutung in der Gegenwart.* Iserlohn: J. Baedecker.

Le Doux, J. (1998). *The emotional brain. The mysterious underpinnings of emotional life.* New York: Touchstone.

Leary, D. (1990). *Metaphors in the history of psychology.* Cambridge: Cambridge University Press.

Leatherdale, W. (1974). *The role of analogy, model and metaphor in science.* New York: American Elsevier.

Libet, B., Freeman, A., & Sutherland, K. (Eds.). (1999). *The volitional brain. Toward a neuroscience of free will.* Exeter: Imprint Academic.

Melnyk, A. (2003). *A physicalist manifesto. Thoroughly modern materialism.* Cambridge: Cambridge University Press.

Meschede, K. (1980). Materialismusstreit. In J. Ritter & K. Gründer (Eds.), *Historisches Wörterbuch der Philosophie, Bd. 5* (pp. 868–869). Schwabe & Co AG: Basel.

Midgley, M. (1994). *Science as salvation. A modern myth and its meaning.* London: Routledge.

Midgley, M. (2003). *The myths we live by.* London: Routledge.

Moleschott, J. (1852/1971). Der Kreislauf des Lebens. In D. Wittich (Ed.), *Schriften zum kleinbürgerlichen Materialismus in Deutschland, Bd 1* (pp. 25–341). Berlin: Akademie.

Moser, P., & Trout, J. D. (Eds.). (1995). *Contemporary materialism: A reader.* London: Routledge.

Moss, J. (1993). *Novelties in the heavens. Rhetoric and science in the Copernican controversy.* Chicago: University of Chicago Press.

Nagel, T. (1974). What is it like to be a bat? *Philosophical Review, 83*(4), 435–450.

Nagel, T. (1986). *The view from nowhere.* Oxford: Oxford University Press.

Nieke, W. (1980). Materialismus. In J. Ritter & K. Gründer (Eds.), *Historisches Wörterbuch der Philosophie, Bd. 5* (pp. 842–850). Basel: Schwabe & Co AG.

Numbers, D. (Ed.). (2009). *Galileo goes to jail and other myths about science and religion.* Cambridge: Harvard University Press.

Onfray, M. (2005). *Traité d'athéologie.* Paris: Grasset & Fasquelle.

Papineau, D. (2002). *Thinking about consciousness.* Oxford: Clarendon.

Penfield, W. (1975). *The mystery of the mind.* Princeton: Princeton University Press.

Popper, K. (1977/1983). Materialism criticized. In K. Popper, & J. Eccles (Eds.), *The self and its brain. An argument for interactionism* (pp. 51–99). New York: Routledge.

Ramachandran, V. S. (2004). *A brief tour of human consciousness.* New York: Pi Press.

Ramachandran, V. S., & Blakeslee, S. (1998). *Phantoms in the brain. Probing the mysteries of the human mind.* New York: Harper Collins.

Searle, J. (1997). *The mystery of consciousness.* New York: New York Review.

Shea, E. (2009). *How the gene got its groove. Figurative language, science, and the rhetoric of the real.* Albany: State University of New York Press.

Stark, R. (2009). *Rhetoric, science, and magic in the seventeenth-century England.* Washington: The Catholic University of America Press.

Tye, M. (2009). *Consciousness revisited. Materialism without phenomenal concepts.* Cambridge: The MIT Press.

Vogt, K. (1847/1971). Physiologische Briefe für Gebildete aller Stände. In D. Wittich (Ed.), *Schriften zum kleinbürgerlichen Materialismus in Deutschland, Bd 1* (p. 1–24). Berlin: Akademie.

Wallace, A. (2000). *The taboo of subjectivity. Toward a new science of consciousness.* Oxford: Oxford University Press.

Wittkau-Horgby, A. (1998). Materialismus. Entstehung und Wirkung in den Wissenschaften des 19. Jahrhunderts. Göttingen: Vandenhoeck und Ruprecht.

Wundt, W. (1889). *System der Philosophie.* Leipzig: Engelmann.

Chapter 2
The Major Objections from Reductive Materialism Against Belief in the Existence of Cartesian Mind–Body Dualism

Robert Almeder

Abstract I discuss five basic objections that materialists often raise to Cartesian Mind-Body Dualism: (1) It is not empirically testable or confirmable; (2) It is in principle testable and confirmable, but unconfirmed; (3) It is testable and confirmable, but has been shown false; (4) It is unnecessary to explain anything; (5) It cannot serve to explain anything. I will show how unsatisfactory all these objections are. If I am right in what I argue the reductionist posture of contemporary materialism against the existence of Cartesian Immaterial Substances as causal agents in explaining human behavior is demonstrably more dogma than anything else. Moreover, the promise of reductive materialism to explain human personality, consciousness, and behavior is unlikely ever to be fulfilled.

Unfortunately, limits on space here prevent a discussion of more positive, empirical research from reincarnation studies and voluntary out-of-body experiments providing solid empirical or scientific evidence affirming both the existence of Cartesian Immaterial Substances and some form of personal survival after biological death. So, even though I will not be able to argue here that we have a well-confirmed scientific belief that there are Cartesian Immaterial causes of human behavior that undermine explanations in usual reductionist efforts, we can at least show why the standard and pervasive objections from reductive materialism fail by way of seeking to show that Cartesian Mind-Body Dualism is false.

R. Almeder (✉)
Georgia State University, Hamilton College (2004–2006), 113 Kings Highway,
Kennebunkport, ME 04046, USA
e-mail: ralmeder@roadrunner.com

A. Moreira-Almeida and F.S. Santos (eds.), *Exploring Frontiers of the Mind-Brain Relationship*, Mindfulness in Behavioral Health, DOI 10.1007/978-1-4614-0647-1_2,
© Springer Science+Business Media, LLC 2012

2.1   Introduction

When confronted with the problem describing human nature or explaining human behavior, most philosophers and scientists think that we do not need, and cannot justify, appealing to the existence of minds or souls as substances distinct from, and in addition to, physical bodies. They believe that only physical objects exist. So, when it comes to minds or mental events proposed as immaterial (or nonphysical) causes of human behavior, they object by insisting that if minds or mental events (such as wishing, believing, intending, or wanting) are not ultimately reducible to, or identical with, some set of brain states, or some complex computational function of brains, or some biological property produced by the brain, then there simply are no minds or mental events or souls. They believe it unscientific to think otherwise. Call their view *reductive materialism* and, because it eliminates by reduction any causally effective "ghosts" in the machine, we can also call it *eliminative materialism*. It has dominated the philosophical and scientific landscape for well over 50 years. Even among those who would otherwise hesitate to characterize themselves as *naturalized epistemologists* (i.e., as people who think that the only legitimately answerable questions are those we can answer by appeal to the methods of testing and confirmation in the natural sciences), there is a strong tendency to accept the view that believing in nonreducible, nonphysical, minds asserted to exist by Plato, Aristotle, the Medievals, Descartes, and other philosophers up to the publication of U.T. Place's influential "Are Mental Events Brain States?" (Place 1956) is just too philosophically and scientifically unjustifiable.

There are, to be sure, voices crying out in the wilderness that reductive materialism may not be true, or even that it is demonstrably false.[1] But it seems clear that the majority of scientists and philosophers of mind continue to regard those voices as the unfortunate legacy of tenaciously entrenched superstition or religion. For that majority, whatever else the mind–body problem may be, it will obviously not extend to the question of whether there are any Cartesian Immaterial Substances that cause certain human behaviors and are somehow irreducible to any physical property, complex or otherwise, chemical or biological, of the brain.

But why do reductive materialists contend that belief in such nonreductive Cartesian immaterial substances is unjustifiable? Alternatively put, what are the basic objections, constituting the core reasons, reductive materialism advances against any Cartesian mind–body dualism, affirming the existence of minds or souls as immaterial causes of human behavior.

---

[1] See, for example, C.D. Broad's *Mind and Its Place in Nature*. London: Routledge and Kegan Paul. 1962. David Chalmer's *The Conscious Mind.*, Oxford University Press. 1990, Joseph Levine's *Purple Haze The Puzzle of Consciousness*, Oxford University Press, 2002 (and reviewed by Terry Horgan in Nous Vol. xl, number 3, Sept. 2006), Richard Swinburne, "Personal Identity: The Dualist Theory" in Personal Identity. Edited by Richard Swinburne and Sydney Shoemaker, Oxford England, Blackwell Publishers, (1999, 3–35), Jaegwon Kim, *Physicalism or Something Near Enough*. Princeton Monographs in Philosophy. 2005. Princeton University Press, Princeton N.J.

2.2   The Five Core Objections to Cartesian Immaterialism

The five core materialist objections to belief in Cartesian Immaterialism are the following:

1. It is not empirically testable or confirmable.
2. It is in principle testable and confirmable, but unconfirmed.
3. It is testable and confirmable, but has been shown false.
4. It is unnecessary to explain anything.
5. It cannot serve to explain anything.

In examining these objections, I shall argue that they are all bad objections, and so natural science has not refuted Cartesian mind–body dualism. Presumably, these are the best objections reductive materialism can offer. Along the way I shall make a modest effort in suggesting more positive reasons for adopting the Cartesian position than the materialist failure to refute it, as we noted above, although lack of space here makes telling this part of the story in full detail a more ambitious undertaking for a future date.

2.2.1   Not Empirically Testable or Confirmable

The *first* and common objection one hears from reductive materialists is that the belief in the existence of Cartesian Immaterial Substances is not empirically testable or confirmable; and so it falls squarely into the domain of philosophy, or religion, or simple superstition. Falling there, they add, is the kiss of death. As we will see when we examine the next objection, however, belief in the existence of Cartesian Immaterial Substance is quite empirically testable and confirmable. But even if, contrary to fact, it were not empirically testable, and if we then ask what is wrong with this issue being a philosophical matter, the typical answer will be that philosophers, unlike scientists, have never really agreed on anything nontrivial. Descartes was right in noting this "scandal of philosophy" but, so this objection continues, in spite of his deepest aspiration to the contrary, Descartes was never quite able to overcome that scandal with his attempt to place philosophy on a firm methodological footing that would allow for something like reliable knowledge about the world. By way of contrast, at least all scientists will, by their method, agree that there are certain *nomic regularities* (causal physical laws) allowing us to predict precisely our sensory experience and certain physical events thereby allowing for our greater adaptability under evolution; and that is the reason why we should insist on verifiability or verification as a necessary condition for any reliable belief about world. The Cartesian can reply, however, that this typical verificationist answer is problematic because it arguably underestimates how much philosophers have agreed upon, and overestimates how much scientists have agreed upon even in the face of theories that allow for very reliable predictions. While we cannot pursue it here at length, philosophers have agreed, for example, that if we take classical logic seriously, philosophical solipsism (the thesis that I alone exist) is indefensible, and most agree that it has been refuted, although certainly not in any laboratory setting or in any experimental or nonexperimental test.[2] Philosophers also generally agree that Aristotle's view that nonhuman animals are not rational because they do not think (because they do not use tools and hence show no capacity to relate means to ends) stands refuted along with Hume's theory of ideas asserting, as it does, that there is no idea that does not derive from some distinct corresponding impression of sense. On the other hand, however well-confirmed a scientific thesis or explanation may be, that in itself does not show it to be objectively true, as if the thesis or explanation as stated and confirmed was forever immune to rejection or serious modification. The history of science is replete with claims that were once well-confirmed and enthusiastically accepted by the scientific community at large only to find later that those same theses were no longer acceptable in the light of new bodies of evidence. Ptolemaic Astronomy (the Geocentric Theory), Absolute Space and Time, The Caloric Theory of Heat, and The Phlogiston Theory of Combustion come readily to mind as suitable examples of such occurrences. In science, as elsewhere, consensus is always desirable and necessary, but it is fallible and evolving with the inevitable increase of evidence and a deep respect for fallibilism. In the long run, science may be no better off than philosophy, even though science can certainly predict our sensory experiences better on any given day. But whether that success counts for more long-term agreement than philosophy on crucial issues seems debatable.

2.2.2 In Principle Testable and Confirmable, but Unconfirmed

The *second* objection one frequently finds among the scientifically-minded is that, contrary to the main point just offered in the first objection, the belief in Cartesian Immaterial Substance, or minds, is indeed empirically testable and confirmable, that it has been tested but not confirmed, and so we have no confirming evidence for it. This objection the philosopher Derek Parfit, among others, offers.

In *Reasons and Persons,* under a heading (#82) entitled *How a Non-Reductionist View Might have Been True,* Parfit (1984) said this:

> Some writers claim that the concept of a Cartesian Ego is unintelligible. I doubt this claim. And I believe that there might have been evidence supporting the Cartesian View.

---

[2] The refutation occurred first when Christine-Ladd Franklin wrote to Bertrand Russell and asserted that she found Russell's defense of Personal Solipsism compelling and that as a result she too was a solipsist. It appears that Russell thought she was serious and that she had not intended to offer a decisive counterexample to any argument favoring Russell's thesis. For a fuller discussion of this and other instances of philosophical agreement see the author's *Harmless Naturalism: An Essay on the Limits of Science and the Nature of Philosophy* (Open Court Publishers, 1998).

There might, for example, have been evidence supporting the belief in reincarnation. One such piece of evidence might be this. A Japanese woman might claim to remember living a life as a Celtic hunter and warrior in the Bronze Age. On the basis of her apparent memories she might make many predictions that could be checked by archaeologists. Thus she might claim to remember having a bronze bracelet, shaped like two fighting dragons. And she might claim that she remembers burying this bracelet beside some particular megalith, just before the battle in which she was killed. Archaeologists might now find just such a bracelet buried in this spot, and their instruments might show that the earth had not been disturbed for at least 2000 years. This Japanese woman might make many other such predictions, all of which are verified.

Suppose next that there are countless other cases in which people alive today claim to remember living certain past lives, and provide similar predictions that are all verified. This becomes true of most of the people in the world's population. If there was enough such evidence, and there was no other way in which we could explain how most of us could know such detailed facts about the distant past, we might have to concede that we have accurate quasi-memories about past lives. We might have to conclude that the Japanese woman has a way of knowing about the life of a Celtic Bronze Age warrior which is like her memory of her own life.

It might next be discovered that there is no physical continuity between the Celtic warrior and the Japanese woman. We might therefore have to abandon the belief that the carrier of memory is the brain. We might have to assume that the cause of these quasi-memories is something purely mental. We might have to assume that there is some purely mental entity, which was in some way involved in the life of the Celtic warrior, and is now in some way involved in the life of the Japanese woman, and which has continued to exist during the thousands of years that have separated the lives of these two people. A Cartesian Ego is just such an entity. If there was sufficient evidence of reincarnation, we might have reason to believe that there really are such entities. And we might then reasonably conclude that such an entity is what each of us really is.

This kind of evidence would not directly support the claim that Cartesian Egos have other special properties in which Cartesians believe. It would not show that the continued existence of these Egos is all-or-nothing. But there might have been evidence to support this claim. There might have been various kinds or degrees of damage to a person's brain which did not in any fundamental way alter this person, while other kinds or degrees of damage seemed to produce a completely new person, in no way psychologically continuous with the original person. Something similar might have been true of the various kinds of mental illness. We might have generally reached the conclusion that these kinds of interference either did nothing at all to destroy psychological continuity, or destroyed it completely. It might have proved impossible to find, or to produce, immediate cases, in which psychological connectedness held to reduced degrees.

Have we good evidence for the belief in reincarnation? And have we evidence to believe that psychological continuity depends chiefly not on the continuity of the brain but on the continuity of some other entity, which either exists unimpaired, or does not exist at all? We do not in fact have the kind of evidence described above. Even if we can understand the concept of a Cartesian Pure Ego, or spiritual substance, we do not have evidence to believe such entities exist. Nor do we have evidence to believe that a person is any other kind of separately existing entity. And we have much evidence both to believe that the carrier of psychological continuity is the brain, and to believe that psychological connectedness could hold to any reduced degree.

I have conceded that the best-known version of the No Reductionist View, which claims that we are Cartesian Egos, may make sense. And I have suggested that, if the facts had been very different, there might have been sufficient evidence to justify belief in this view (p. 227–228).

Parfit's argument avoids the easy dogmatism that comes of casually affirming that belief in the existence of Cartesian Immaterial Substance is obviously unintelligible, or pure religion, or superstition. Doubtless, from some point of view such an assertion is unintelligible, but whether that point of view is in fact defensible, or possibly adopted as philosophical dogma, would still be an open question. Rather than take the dogmatic stance, Parfit seeks instead to satisfy the question "What should you take as solid empirical evidence that Cartesian Immaterial Substances exist?" Along with Ayer (1956), Parfit affirmed that the thesis is in fact empirically testable under a minimalist construal of reincarnation because, he says, the reincarnation hypothesis makes predictions or has test implications at the sensory level, as long as we accept as a necessary condition for personal identity that one have systemic memories that nobody else could have.

After all, if the person beside you professed to having memories that only Julius Caesar could have had, and indeed had a number of such memories, and if we could find no other equally plausible way to explain how he got the memories of Julius Caesar, then, assuming also that we had a large number of similar cases, you would be stuck with the claim that we have solid evidence here that the person beside you is Julius Caesar. Indeed if s/he really is Julius Caesar in a different body, then s/he should have empirically confirmable memories that only Julius Caesar could have had. He might, for example, tell you that he had a twin brother, Caius, who for certain reasons never left the family farm, and that he, Julius, buried at a specific location beside the Rubicon river just before crossing it in 49 b.c. a sum of 500 newly minted gold coins and a personal note leaving the money to Caius, just in case things did not go well in Rome the following week. He also tells you he instructed one of his soldiers, Cratylus, to go to his family secretly and inform them of the whereabouts of the buried gold coins for Caius. Such memory claims would indeed be empirically confirmable in the way suggested by Parfit. Certainly, too, there is no current record anywhere that Caesar had a twin brother, or that he, Caesar, indeed buried that sum of money with a note leaving the money to his twin brother Caius. Assume then that we find the place where the money is supposedly buried, excavate carefully the ground, assured by paleontologists that the ground has not been disturbed at that site since 49 b.c., and find the minted gold coins and the note (authenticated by several distinguished graphologist) in the handwriting of Julius Caesar. Suppose further that the person beside you continues and tells you many other similar memories he has, and suppose they are all confirmed in much the same way. How would we explain his having these confirmed memories that *only* Julius Caesar could have had? What would be the best available and nonarbitrary explanation for this person having these memories if it is not that this person beside you is indeed Julius Caesar in a new body? Certainly most people would take such evidence as confirmatory of this person being Julius Caesar in a new body and sitting beside you. That is because most people instinctively believe the memory theory of personal identity, which neither Parfit nor

Ayer questioned.[3] Further, if we could not come up with an alternative empirically testable hypotheses that would produce the same effects without our having to believe that the person beside you is Julius Caesar, and if there were many other cases like this, then we would have little choice but to accept that at least this person beside you is reincarnated, by some causal mechanism we know not what, and for some reason we know not why.

Along with Parfit and Ayer, of course, if we do not abandon the having of systemic and unique memories as at least a necessary condition for personal identity, not to say essential, and if something like the above came to pass very often, we would need to change dramatically what we mean by memory, because we could not define memory in terms of some biological product of the brain, or some neural network, or any describable biochemical property or complex set of neurobiological properties, ultimately defined in terms of atoms and molecules that are governed by the laws of physics at some fundamental level and cease to exist with the death of the brain. Those things die with the brain. But this person beside you has the confirmed memories and not the "quasi memories" of Julius Caesar, and if Julius Caesar's memory was identifiable with the above stated properties or biological properties produced by the brain, obviously this person could not have the memories s/he does have.

---

[3] The major objection to any memory theory of identity was offered briefly by Thomas Reid and later at greater length by Bernard Williams in "Personal Identity and Individuation" (*Proceedings of the Aristotelian Society* 57, 1956–1957). The objection consists in a thought experiment in which a fellow named Charles turns up claiming to be Guy Fawkes. All the events he claims to have witnessed back as Fawkes in the sixteenth century and the actions he claimed to have done point unanimously to a life-history of some one person in the past – Guy Fawkes. All Charles' memory claims can be checked to fit the life of Fawks and some few that cannot will be plausible, and provide explanations of unexplained facts. And so by the memory theory of personal identity, Charles is Guy Fawkes in a new body. Williams asks us now to imagine that another person, Robert, who turns up and satisfies the memory criteria for being Guy Fawkes equally well. We cannot say they are both identical with Guy Fawkes, because if they were they would be identical to each other which they are not because they currently live different lives different thoughts and feelings from each other. So, Williams concludes that apparent memory cannot constitute personal identity. This basic objection has convinced the majority of writers that something more like bodily continuity than memory would count for personal identity. As I see it however, the counterexample does not work. If Robert did show up satisfying the memory criteria for being Guy Fawkes, that would be an empirical disproof of the memory criteria for identity. The attractiveness of the memory criteria for personal identity is that it is in fact empirically falsifiable just in case somebody other than Charles was to show up with the same memories of Guy Fawkes. The fact that we can imagine the empirical events that would falsify the memory theory of identity is not a logical refutation of the theory, but rather a statement of conditions that would be sufficient to empirically refute the theory. I think we would all agree that if Robert had the same memories as Guy Fawkes while Charles has them also, the memory theory of personal identity would stand refuted. But that has not happened just yet and so the memory theory cannot be simply dismissed by appeal to what we would accept as empirical evidence for the falsity of the memory theory of personal identity. I know of no other persuasive counterexample to the memory theory of identity and, in fact, given that the thesis is empirically falsifiable it seems strange to try to offer a counterexample, as if it were a matter of taking the theory as an instance of definition by way of appeal to ordinary usage, rather than an empirically falsifiable thesis.

So, belief in the existence of Cartesian Immaterial Substances is an empirically confirmable hypothesis. That conclusion in itself should be big news, and one can only wonder why so little has been said about it, given the general assumption so widely adopted that whether anything like a Cartesian Mind exists is a question of Metaphysics (in the pejorative sense) and not something that is empirically confirmable or testable in natural science.[4]

Anyway, Parfit hastens to add that while belief in reincarnation and, by implication, in Cartesian Immaterial Substance, is certainly an empirically testable and confirmable thesis, we do not in fact have any such evidence for believing in reincarnation. Although the above cited Parfit text is not as clear as one might wish, he seems to assert that the thesis has been tested, and that we never got the confirming evidence to warrant acceptance of Cartesian Immaterial Substance, and so we would have no rational justification for accepting the thesis. This conclusion emerges because Parfit, unlike Ayer, thinks a necessary condition for accepting the thesis would require most of the current population to have such confirmed memories before we could say of anyone in particular that s/he is a reincarnated person. That requirement is an unusually and arbitrarily strong requirement, rather than say a requirement to the effect that there be over time a large number of such cases, enough to establish the nonanecdotal nature of the evidence offered. If that is so, and arguably it is, then Parfit's position would be that science has at least indirectly refuted the Cartesian position by indirectly refuting the thesis of reincarnation, which he apparently takes to be the only hypothesis under which Cartesian Dualism is empirically testable. Apart from that, he offers us no help in what it would take to *disconfirm* either reincarnation or dualism under some other hypothesis.

Nor should we forget, incidentally, that Ayer (1956: 193) in the course of arguing for memory as the criterion for personal identity, argued that if the man sitting bedside you had the memories of Caesar Augustus, better yet had memories that *only* Caesar Augustus could have had, and such memories were confirmed, then we would need to say that the man beside you is indeed Caesar Augustus in a different body, unless we could find some way to confirm the belief that one could have the memories that only Caesar Augustus could have without being Caesar Augustus. Ayer, like Parfit, had no hesitation in accepting the view that the existence of minds is an empirical hypothesis testable under the hypothesis of reincarnation. In fact, however, Ayer did not believe ostensibly in reincarnation, but used it rather as a thought experiment to drive home what we would say if the evidence for reincarnation actually obtained. Clearly he took the existence of Cartesian minds, by implication, to be an empirically testable thesis as long as one accepts *the memory theory of personal identity*. This conclusion was orthogonal to his well-published earlier view that empirical hypotheses that are central to the sciences are not at the core of

---

[4] For other ways one could empirically test for the existence of Cartesian Immaterial Substances, see Chap. 3 and the discussion of the Osis-McCormick Experiment with reference to voluntary out-of-body experiences pp. 167–202 of the author's *Death and Personal Survival* 1992. Littlefield Adams, Quality Paperback.

philosophy (Ayer 1994).[5] One would have expected that the mind–body problem is at the core of philosophy, but his granting it empirical status as a testable empirical hypothesis would place it squarely inside natural science.

In the end what seems objectionable in Parfit's claim that we do not have the required evidence for justified belief in reincarnation is simply that he lays down an impossibly strong requirement, namely that most of the current population have at the same time empirically confirmed past life memories that only the former person could have had. Fortunately, Ayer made no such a demand rather than that there simply be many other past similar cases of confirmed memories. Let us turn briefly to John Searle's position which is offered as evidence for the claim that science has shown that traditional or Cartesian mind–body dualism is false, and not simply that we do not have enough evidence for it.

2.2.3   Testable and Confirmable, but Shown to Be False

The *third* objection one will see sooner or later is that the belief in Cartesian Immaterial Substance is indeed empirically testable but science *shows* that souls, or Cartesian Immaterial Substances, cannot exist because contemporary science *shows* that consciousness, or any mental state whatever, at least as traditionally conceived, cannot exist after the death of the brain. In *The Rediscovery of Mind*, after asserting that all mental events are biological phenomena, Searle (1992) goes on to say of them:

> They are as much the result of biological evolution as any other phenotype. *Consciousness, in short, is a biological feature of human and certain animal brains. It is caused by neurobiological processes and is as much a part of the natural biological order as any other biological features such as photosynthesis, digestion, or mitosis.* This principle is the first stage in understanding the place of consciousness within our world-view. The thesis of this chapter so far has been that once you see that atomic and evolutionary theories are central to the contemporary scientific worldview, then consciousness falls into place naturally as an evolved phenotypical trait of certain types of organisms with highly developed nervous systems. I am not in this chapter concerned to defend this worldview. Indeed, many thinkers whose opinions I respect, most notably Wittgenstein, regard it as in varying degrees repulsive, degrading and disgusting. It seems to them to allow no place --or at most a subsidiary place—for religion, art, mysticism, and "spiritual values" generally. But like it or not, it is the worldview we have. Given what we know about the details of the world—about such things as the position of elements in the periodic table, the number of chromosomes in the cells of different species, and the nature of the chemical bond---this world view is not an option. It is not simply up for grabs along with a number of competing worldviews. Our problem is not that we have somehow failed to come up with a convincing proof of the existence of God or that the hypothesis of an afterlife remains in serious doubt, it is rather that in our deepest reflections we cannot take such opinions seriously. When we encounter

[5] A. J. Ayer, "On Making Philosophy Intelligible" in *Metaphysics and Common Sense*. Jones and Bartlett. 1994. 1–19; *The Problem of Knowledge*. 1956. Penguin Books, Pelican Paperback, 187–20.

people who claim to believe such things, we may envy them the comfort and security they claim to derive from these beliefs, but at bottom we remain convinced that either they have not heard the good news or they are in the grip of faith. We remain convinced that somehow they must separate their minds into separate compartments to believe such things. When I lectured on the mind-body problem in India and was assured by several members of my audience that my views must be mistaken, because they personally had existed in their earlier lives as frogs or elephants, etc., I did not think "Here is evidence for an alternative world view" or even "Who knows, perhaps they are right" And my insensitivity was much more than mere cultural provincialism: Given what I *know* about how the world works, I could not regard their views as serious candidates for truth.

And once you accept our world view the only obstacle to granting consciousness its status as a biological feature of organisms is the outmoded dualistic/materialistic assumption that the "mental" character of consciousness makes it impossible for it to be a physical property. (p.90-91)....Anyone who has had even a modicum of scientific education after about 1920 should find nothing at all contentious or controversial in what I have just said. It is worth emphasizing also that all of this has been said without any of the traditional Cartesian categories. There has been no question of dualism, monism, materialism, or anything of the sort. Furthermore there has been no question of "naturalizing consciousness"; it already is completely natural. Consciousness, to repeat, is a natural biological phenomenon. The exclusion of consciousness from the natural world was a useful heuristic device in the seventeenth century, because it enabled scientists to concentrate on phenomena that were measurable, objective and meaningless, that is, free of intentionality. But the exclusion was based on a falsehood. It was based on the false belief that consciousness was not part of the natural world. That single falsehood, more than anything else, more even than the sheer difficulty of studying consciousness with our available scientific tools has prevented us from arriving at an understanding of consciousness. (p.93. for an essentially identical assertion, see Searle 2004)

Searle's argument, then, for the claim that consciousness exists as a *biological* product of the brain, secreted by the brain in the same way a hormone is secreted by a gland, is then simply that that is the only position consistent with a naturalistic world view in which what is known about the world is what we can get under the method of testing and confirmation in the natural sciences as we have come to know them. He is quick to add, of course, that he is not interested in defending such a worldview. He simply accepts it as obvious that it is our worldview, and asserts that that fact in itself should be sufficient reason for the rest of us to accept it, and to make our philosophical explanations consistent with it (Searle 2004: 101). So he urges that those who would affirm the existence of consciousness as a Cartesian Immaterial Substance, and thereby reject the biological nature of consciousness, disagree with our world-view; and they thereby do so either because they are in the grip of religion or just have not yet heard the good news that science and the scientific world view is all we have when it comes to knowing anything about this world. They either know nothing about science, or they are superstitious.

In fact, by way of criticism of Searle's position, it is quite possible to accept *a* scientific world view, in any of the various ways Searle might be inclined to define or characterize it, and still, without being superstitious or essentially ignorant of science, reject Searle's biological construal of consciousness, simply because his position is purely philosophical and not in fact established in natural science. His position on the biological nature of consciousness contradicts his stated worldview. After all, where in the scientific literature, biological, neurobiological, or otherwise, is it established either by observation or by the methods of testing and experiment, that consciousness is a biological property secreted by the brain in the same way a gland secretes a hormone? Better yet, where in the history of science has it been established that consciousness exists, but cannot be a substance very much unlike any substance we ordinarily deal with in contemporary physics or biology? In short, there is no scientifically well-confirmed (much less robustly confirmed) belief within science that consciousness is a biological product of the brain. We do not see the brain secrete consciousness in the same way we see a gland secrete a hormone. Consciousness is nothing like a hormone.

When this last objection is noted, the Searlean materialist's fall-back position is that nevertheless the biological construal of consciousness is the only position consistent with our scientific world view. Supposing indeed that to be the case, where was that world-view established as a truth or a robustly confirmed hypothesis in science? Besides that, what exactly does Searle mean by "our scientific world view?" Well, of course, he said above that adopting the scientific world view is just another way of saying that in the interest of attaining human knowledge we need to *naturalize* everything and take the methods of science as the only way to attain human knowledge. But that is arguably a bit too vague because the concept of a scientific world view admits of no fewer than three logically distinct characterizations[6] and, depending on which characterization one chooses, one may or may not have a justification for adopting a scientific world view; and one can argue that the only viable characterization of "scientific world view" that is harmless, is the one that leaves it an open question as to what the nature of consciousness might turn out to be.

Finally, Searle apparently believes that simply because we have adopted a scientific world-view, (in some sense suitably explicated) then whether anybody likes it or not, that is a good reason for adopting it. Given all this, and when all the appeals to obviousness are done, the ultimately nagging question is why should anybody take seriously the *biologizing* of consciousness as something warranted in science or even as something warranted in terms of accepting a scientific world-view? Searle's claim that science has *shown* that consciousness, like any mental state is a biological property of the brain and hence dies with the death of the brain, is by no means as obvious as he contends. Indeed, it is false that science has shown as much. Nobody, as we remarked earlier, has yet seen consciousness secreted by the brain in the way one can see a hormone secreted by a gland. It is also false that science has

---

[6] There are no fewer than three logically distinct forms of naturalized epistemology: (a) The only legitimately answerable questions about this world are those we can answer by appeal to the methods of testing and confirmation in the natural sciences, and the only correct answers we have are those provided by natural science (replacement thesis); (b) There are legitimately answerable questions outside of natural science, but whether anybody knows anything or not is an empirical or scientific question (transformational thesis); and (c) The method of the natural sciences is the only reliable method for acquiring a public understanding of the nature of the observed regularities and properties of the physical world.(harmless thesis). For a full discussion of all three and an endorsement of (3), see the author's *Harmless Naturalism: The Limits of Science and the Nature of Philosophy.* (Open Court Publishers, Chicago, Illinois). 1998. p b.

shown that consciousness cannot be some sort of Cartesian Immaterial Substance irreducible radically to any property of the brain. And even if Searle's biologizing of consciousness and all other mental states were the only position consistent with accepting our scientific world-view, Searle's refusal to defend such a world view reveals at best a lack of understanding of those arguments already in the literature to the effect that naturalizing everything (under either the replacement thesis or the transformation thesis) is fraught with difficulties, and at worst an elementary *ad populum.*[7]

2.2.4 *Unnecessary to Explain Anything*

The *fourth* core objection is that we simply do not need Cartesian Immaterial Substance to explain anything at all. We only need physical laws and physical objects to explain and predict all of human behavior, and even if we cannot now predict all of human behavior, at least it is something we can do in principle. This common objection feeds upon the traditional *Principle of Parsimony* which asserts that the only justification we have for believing in the existence of anything is that the belief explains something we could not otherwise explain equally well without that belief. Bypassing certain questions about what would count as an adequate theory for the explanation of human behavior, and whether the ability to predict human behavior in itself would count for such an explanation, this objection is, more than anything else, a challenge to the Cartesian dualist to come up with good reasons for supposing that we need something more, or that there is something fundamentally wrong with commonly proposed explanations put forth by reductive materialists to explain human behavior. Here the trench warfare begins.

Take for example, the problem of consciousness. Consciousness certainly does not seem to be a property like any other physical property. Everybody admits that it exists, but being generally aware of things is not like any other physical property we know about, or whose existence can be directly or indirectly inferred from observation of other physical properties. Reductive Materialists, however, will generally argue that being in a particular brain state just *is* being conscious; certain describable

---

[7] There are other problems with Searle's proposed solution to the mind–body problem. He asserts, for example, that materialism and traditional dualism are false, that is, it is false that only material objects exist (because there are mental states) and it is false that traditional dualism is true (because that implies that there are substances not reducible to physical objects). For Searle, the trick is to realize that there are mental states and consciousness, but that they are in fact material or biological states of the system produced by the brain. (RM 15 and HPR Spring 2004. 110–113) But arguably that just is classical materialism. How this differs from the original form of eliminative materialism offered by Rorty in 1965, (Review of Metaphysics. Vol. 19) is difficult to fathom. Searle's view is fundamentally that there are mental events but they are material events, and this is clear if we can see that there is no disjunct between the mental and the material. Once we get over that hump we will see that the original mind–body problem was generated by a bad definition of the mental and the material, one that made the mental and the material mutually exclusive.

neuro-biological activities always occur when consciousness is present, but are not there when consciousness is not present (Dennett 1992). Note also here that Dennett's *Consciousness Explained* is in fact no different from that form of eliminative materialism that falls under the contingent identity thesis. Less popularly, as we saw above in the work of John Searle, other reductive materialists will argue that *being conscious* does not reduce simply to being in a particular brain state describable simply in neuro-chemistry, but rather reduces, a la Searle, to *being a biological property produced or secreted by the brain.* We need not repeat the above reasons why the latter form of reductionism seems so unsatisfactory. But the idea that consciousness just *is* being in a particular neuro-biological state, complex or otherwise, which *is* the awareness we experience, as materialists say, does not seem to the Cartesian Dualist to be any more empirically confirmed than alternative explanations such as the position offered by John Searle, or even that offered by Cartesians who might urge that consciousness is neither a physical property nor any other empirically describable state of the brain; but rather that when consciousness is present certain parts of the brain light up, as it were, and would not light up otherwise because that is the way in which consciousness causes the brain to do the work it does in producing various human behaviors such as believing, desiring, intending, remembering, loving, hating, and knowing. In addition, as we shall see soon, there are other human behaviors that we cannot explain simply by some appeal to causal brain states as either causative or constitutive of such behavior.

2.2.5   It Cannot Serve to Explain Anything

Our *fifth reductionist* objection to believing in Cartesian immaterial substances is that such substances, even if they existed, could not function as causes of anything in the world, and so the belief in them could have no explanatory power in principal for anything and especially when it comes to explaining human behavior. This objection has a long history. It works on the principle that anything that will be a cause in our explaining observable human behavior will need to function by way of conveying kinetic energy to another object; otherwise there would be no explanation for the human behavior that occurs because we would not be able to predict the behavior under the cause. If Cartesian Immaterial Substance could be a cause, then something could occur without the transfer of kinetic energy and, so this anti-Cartesian objection goes, that would violate *The Principle of Conservation of Energy*, as it would allow for the overall increase of energy in the universe, just as if physical events could cause mental events, there would be an overall decrease in energy in the universe.

However, seductive this objection to belief in the existence of Cartesian Immaterial Substances as causally productive of human behavior, it suffers from at least one fundamental flaw. The concept of causality to which the materialist appeals begs the question in favor of his position that only physical objects exist, because he defines a cause not simply as that object whose efficient action brings about a change in another object, but rather as that object by whose conveyance of kinetic energy brings about another proportional and predictable change in the observable properties of the other object. When one defines causality in this way under the rubric of operationalizing basic concepts in science, the definition *assumes* that causality is a relationship between physical objects and is determinably present only when there is a transfer of kinetic energy in the way understood by traditional physics. This assumption begs the question in favor of a concept of causality that obtains only between physical objects as we know them, and thus begs the question against any basic causal relation between a physical object and a Cartesian Immaterial Substance. It begs the question in favor of mechanistic explanations of human behavior to the extent that concept of causality is also implied in mechanistic explanations. The anti-Cartesian would respond predictably, then, that the Cartesian dualist is unfortunately asking us to take seriously the proposition that we cannot have explanations of human behavior (however, much success we might have in predicting human behavior) in the natural sciences as we know them. Materialists often think that this response closes the debate, because it is hard to take seriously anybody who thinks that natural science cannot provide us with any explanations of human behavior.

But what if Cartesian dualists are willing to accept that particular conclusion and relegate natural science to securing causal explanations among physical objects requiring the transfer of kinetic energy, and then reserve explanations of human behavior for a different type of causal interaction, a basically primitive one wherein there is in fact a transfer of efficient energy between mental and physical objects but not to be understood in terms of a transfer of kinetic energy between two typically observable physical objects? Science, as we currently understand it, may not be able to provide scientifically mature causal explanations of human behavior under this model, but it might still be able to predict a good deal of human behavior from many antecedent statistical correlations. Just as the unpredictable at any moment can and does occur but still has a cause, the predictable can and does occur without our being able to describe the cause in terms of a transfer of kinetic energy from one physical object to another. But, of course, at this point, the anti-Cartesian materialist may quite possibly continue to urge that we cannot then make any scientific sense of a causal relationship between the physical and the nonphysical, and that the supposition to the contrary is somehow incoherent.

On this last point, and in an effort to establish the claim that it is neither logically impossible nor factually impossible that there can be a causal relations between physical objects and Cartesian Immaterial Substances, Broad (1962) once asked us to reflect on our own behaviors and experience of causality.[8] When I raise my arm, for example, just after saying "I will now raise my arm" we usually explain the arm going up by saying he raised his arm because he wanted to raise his arm. Or he raised his arm because he intended to raise his arm. The anti-Cartesian materialist will not deny such explanations, but he will add that wanting, or intending, must be construed as causal agents identical to certain brain states that cause the arm to go

---

[8] *Mind and its Place in Nature*. London: Routledge and Kegan Paul, 1962.

up; its just a case of brain–body interaction and there is nothing particularly myste-rious here. C.D. Broad, however, might have pressed the issue further. Why, he could ask, does my arm go up *just after* I say it will, or, better yet, why does the brain cause the arm to go up at that time and at no other time, than just after I say "I will now raise my arm?" What caused the brain to function as such an effective cause at that point and not before or after? What activates the brain as a causal agent? If my arm went up autonomically, as a result of some neurological glitch, twitch, or of some sort of chemical imbalance, we would not say I raised my arm. What causes the brain to be in precisely the position it needs to be in order to cause the arm to go up at precisely that time when I say "I will now raise my arm?"…and when I do not intend to raise it, or do not want to raise it, why is the brain state not then causing the arm to go up? If the answer here is that there is some other com-plex, or even some simple brain state that is at work to cause the brain to raise the arm, then the next question will be why does that particular cause of the brain activ-ity occur at that time and at no other time? And so we have to go to an infinite regress to explain why my arm is caused to go up by the brain at precisely that time when I say I will raise it and do raise it. This seems problematic for any proposed causal explanation of behavior in terms of intentions and wants that are presump-tively reducible to brain or neuro-biological states. For the Cartesian such a problem leads to the view that there are Cartesian Immaterial Substances causally responsi-ble for human behavior. To be told repeatedly, however, that there could be no such causes of human behavior because we could not understand in science how they work, is simply begging the question against their existence when there is good reason to think that the reductivist thesis fails to explain something as simple and as important as intentional acts such as deliberately raising one's arm at a particular time. Our not knowing in natural science how such causes work does not imply that there are no such causes, but only that we cannot understand them at the moment if we construe them as mechanisms that require a transfer of kinetic energy between two fundamentally physical objects as we ordinarily understand them operationally in scientific contexts. Doubtless, the Cartesian Dualist will claim we are dealing here with some primitive and fundamentally different kind of causation between two different types of objects, although mental events and physical events will obvi-ously need to share something in common for them to be enough alike for there to be any causal interaction at all.

2.3   Conclusion

In his excellent book *Purple Haze: The Puzzle of Consciousness* Levine (2002) is right to say that the antinomy in discussions on the problem of consciousness is that consciousness seems to be so basically irreducible to some interesting physical or material property and yet at the same time we feel the need for causal explanations which belief in irreducible consciousness undermines. This tension goes to the heart of the mind–body problem. The Cartesian Mind–Body Dualist cannot help but be attentive to that antinomy. But if what we have argued above is persuasive by way of funding a rejection to all core objections to Cartesian Mind–Body dualism, we do not need to give up the thesis that Cartesian Immaterial Substances exist and are causes of human behavior. We only need to give up the idea that we can provide causal explanations of human behavior simply in terms of causes understood mechanistically or in terms of the transfer of kinetic energy as we usually understand it.

For lack of space, there is nothing said here about what I have argued elsewhere to the effect that there is indeed commanding empirical evidence that some form of Cartesian Dualism is correct in terms of the strong empirical evidence for some *minimalist* form of reincarnation, and in terms of repeated, and repeatable, case studies of voluntary out-of-body experiences. Reductive Materialists of different stripes tend to ignore that evidence for the alleged reason that it is not scientific and for the further reason (among others) that there is no scientific evidence for any such Cartesian Dualism. On this see Almeder (1992), *Death and Personal Survival: The Evidence for Life After Death*; and also the three-essay exchange between Almeder (2001) and Hales (2001a, b) on reincarnation and science in *Philosophia* and reprinted in Hales and Lowe (2006).

There may be important why-questions about human behavior, questions we cannot answer by appeal to the methods of testing and confirmation in natural science as we currently understand them. If that is true, it raises serious further questions about the science of psychology, and whether it is really explaining human behavior, rather than using statistical correlations to successfully predict a good deal of human behavior. The latter of course is profoundly important and useful without our having to claim we are therein advancing causal explanations of human behavior.

References

Almeder, R. (1992). *Death and personal survival: The evidence for life after death*. Lanham: Littlefield Adams Quality Paperback.

Almeder, R. (1998). *Harmless naturalism: The limits of science and the nature of philosophy*. Chicago: Open Court Publishers.

Almeder, R. (2001). On reincarnation: A reply to Hales. *Philosophia, 28*(1–4), 347–358.

Ayer, J. (1956). *The problem of knowledge*. Baltimore: Penguin Books.

Ayer, J. (1994). On making philosophy intelligible. In Metaphysics and common sense. Jones and Bartlett Publishers.

Broad, C. D. (1962). *Mind and its place in nature*. London: Routledge and Keegan Paul.

Dennett, D. (1992). *Consciousness explained*. Boston: Little Brown and Company.

Hales, S. D. (2001a). Evidence and the afterlife. *Philosophia, 28*(1–4), 335–346.

Hales, S. D. (2001b). Reincarnation redux. *Philosophia, 28*(1–4), 359–367.

Hales, S. D., & Lowe, S. C. (2006). *Delight in thinking: An introduction to philosophy*. New York: McGraw-Hill.

Levine, J. (2002). *The Purple Haze: The problem of consciousness*. Oxford: Oxford University Press.

Parfit, D. (1984). *Reasons and persons* (pp. 227–228). Oxford University Press: Oxford.

Place, U. T. (1956). Are mental events brain states? *British Journal of Psychology, 47*, 44–50.

Searle, J. (1992). *The rediscovery of mind*. Cambridge: MIT Press.

Searle, J. (2004). *Harvard philosophical review* (pp. 111–113). Harvard University Press: Cambridge.

Swinburne, R. (1999). Personal identity: The dualist theory. In R. Swinburne & S. Shoemaker (Eds.), *Personal identity* (pp. 3–35). Blackwell Publishers: Oxford.

Chapter 3
Psychic Phenomena and the Mind–Body Problem: Historical Notes on a Neglected Conceptual Tradition

Carlos S. Alvarado

Abstract Although there is a long tradition of philosophical and historical discussions of the mind–body problem, most of them make no mention of psychic phenomena as having implications for such an issue. This chapter is an overview of selected writings published in the nineteenth and twentieth centuries literatures of mesmerism, spiritualism, and psychical research whose authors have discussed apparitions, telepathy, clairvoyance, out-of-body experiences, and other parapsychological phenomena as evidence for the existence of a principle separate from the body and responsible for consciousness. Some writers discussed here include individuals from different time periods. Among them are John Beloff, J.C. Colquhoun, Carl du Prel, Camille Flammarion, J.H. Jung-Stilling, Frederic W.H. Myers, and J.B. Rhine. Rather than defend the validity of their position, my purpose is to document the existence of an intellectual and conceptual tradition that has been neglected by philosophers and others in their discussions of the mind–body problem and aspects of its history.

> "The paramount importance of psychical research lies in its demonstration of the fact that the physical plane is not the whole of Nature" English physicist William F. Barrett (1918, p. 179)

3.1 Introduction

In his book *Body and Mind*, the British psychologist William McDougall (1871–1938) referred to the "psychophysical-problem" as "the problem of the relation between body and mind" (McDougall 1911, p. vii). Echoing many before him,

C.S. Alvarado, PhD (✉)
Atlantic University, 215 67th Street, Virginia Beach, VA 23451, USA
e-mail: carlos.alvarado@atlanticuniv.edu

A. Moreira-Almeida and F.S. Santos (eds.), *Exploring Frontiers of the Mind-Brain Relationship*, Mindfulness in Behavioral Health, DOI 10.1007/978-1-4614-0647-1_3, © Springer Science+Business Media, LLC 2012

McDougall believed that "any answer to this question must have some bearing upon the fundamental doctrines of religion and upon our estimate of man's position and destiny in the world" (p. vii). Indeed by the time McDougall was writing there was already a long history of speculations and writings about the mind and its various postulated relations to the body (Crane and Patterson 2000; MacDonald 2003; McDougall 1911; Wright and Potter 2000). This included, among others, ideas describing and conceptualizing the mind as an epiphenomenon or as a principle independent of bodily functions. The latter notion and the topic of interest of my chapter was what McDougall referred to in his book as the idea that "all, or some, of those manifestations of life and mind which distinguish the living man from the corpse and from inorganic bodies are due to the operation within him of something which is of a nature different from that of the body, an animating principle generally, but not necessarily or always, conceived as an immaterial and individual being or soul" (p. viii).

In this chapter, I will not attempt to identify the various solutions of mind–body relation proposed over time. Instead my purpose is to focus on a generally neglected intellectual tradition whose representatives defended the existence of a principle independent from the body based on the existence of so-called psychic, supernormal, or parapsychological phenomena such as apparitions, mediumship, clairvoyance, and telepathy. Furthermore, my focus will be on the nineteenth and twentieth-century writings.

3.2   Psychic Phenomena

Such phenomena came into the modern Western age through a variety of movements, among them mesmerism and spiritualism (Dingwall 1967–1968; Podmore 1902), not to mention many other beliefs and practices coming from the past (Goodrick-Clarke 2008; Watkins 2007). Many books presented accounts of psychic phenomena that challenged materialist views in which the mind (or the spirit or soul, depending on the formula of each author) could transcend the physical body and thus show its independence from the body. Examples are *The Night-Side of Nature* (Crowe 1848), *Le livre des esprits* (Kardec 1857), and *Die mystischen Erscheinungen der menschlichen Natur* (Perty 1861). These were early efforts that supplemented the existing dualistic philosophical speculations about the mind or the spirit.

Although occurrences referred to as psychic phenomena have been recorded from ancient times (Figuier 1860; Inglis 1992), more systematic studies have taken place in more modern eras. In addition to mesmerism (eighteenth and nineteenth centuries) and spiritualism (nineteenth and twentieth centuries), psychical research developed in the last quarter of the nineteenth-century. In 1882, the Society for Psychical Research (SPR) was founded in London with the purpose of presenting "an organised and systematic attempt to investigate that large group of debatable phenomena designated by such terms as mesmeric, psychical, and Spiritualistic" (Objects of the Society 1882, p. 3). While the SPR had several spiritualists as members who contributed to the development of the Society what made the organization different was that they also had many prominent academic members. This included Cambridge University scholars such as philosopher Henry Sidgwick (1838–1900) and classicist and poet Frederic W.H. Myers (1843–1901). Furthermore, other eminent individuals associated with the Society included physicists William Barrett (1844–1925) and Balfour Stewart (1828–1887), and politician Arthur Balfour (1848–1930), who later became prime minister. In later years, many eminent men became presidents of the SPR. Examples include American psychologist and philosopher William James (1842–1910), English chemist and physicist William Crookes (1832–1919), English physicist Oliver Lodge (1851–1940), French physiologist Charles Richet (1850–1935), and French philosopher Henri Bergson (1859–1941). Together with the study of spontaneous telepathy, members of the SPR studied telepathy through experiments, and analyzed cases of haunted houses, mediums, and apparitions of deceased individuals (Gauld 1968). The Society also sponsored studies and discussions about dissociative phenomena and the subconscious mind (Alvarado 2002). But there were also studies coming from other countries, among them France and Germany (Brower 2010; Wolffram 2009).

Some of the individuals involved in these studies supported the notion that the mind was a principle separate from the body. One of the leaders of the SPR, the above-mentioned classical scholar Frederic W.H. Myers, saw their work as follows:

> First … we adopt the ancient belief … that the world as a whole, spiritual and material together, has in some way a systematic unity; and on this we base the novel presumption that there should be a unity of method in the investigation of all fact. We hold therefore that the attitude, the habits of mind, the methods, by aid of which physical science has grown deep and wide, should be applied also to the spiritual world. We endeavour to approach the problems of that world by careful collection, scrutiny, testing, of particular facts; and we account no unexplained fact too trivial for our attention (Myers 1900, p. 117).

The issue was expressed in the work of William James, who considered psychical research a valid empirical approach for the study of the mind. As seen in James' (1890, 1909) work with medium Leonora E. Piper (1857–1950), and in other work, James believed that empirical studies of psychic phenomena were important to understand consciousness. This led him to become involved with the work of the SPR, of which he was president in 1896 and with the founding and early investigations of the American Society for Psychical Research (Alvarado and Krippner 2010; Taylor 1996).

In his book *Human Immortality* James (1898) discussed "transmission" and "production" ideas to account for consciousness. This referred to the independence of the mind from the body and to epiphenomenalism, respectively. The first was the assumption that the mind manifested through the nervous system but was an independent principle, while the second was the idea that the mind was produced by the nervous system. As he wrote about these ideas and psychic phenomena:

> A medium … will show knowledge of his sitter's private affairs which it seems impossible he should have acquired through sight or hearing, or inference therefrom. Or you will have an

apparition of some one who is now dying hundreds of miles away. On the production - theory one does not see from what sensations such odd bits of knowledge are produced. On the transmission-theory, they don't have to be 'produced,' – they exist ready-made in the transcendental world, and all that is needed is an abnormal lowering of the brain-threshold to let them through (James 1898, p. 26).

Following James, psychical researcher Hereward Carrington (1880–1958) stated that psychic phenomena "are extremely difficult, if not impossible, to explain and classify on the 'production theory'" (Carrington 1905, p. 46).

But regardless of such ideas in this section I would like to focus on two aspects: First, the prevalent past skepticism about these ideas and second, recent trends in historical writings about psychological topics.

Although some individuals believed in the existence of psychic phenomena, and used such manifestations to promote belief in a spiritual nature in humankind, such interpretations were not shared by many, and certainly not by the majority of the scientific establishment. Readers of this chapter should be aware that this topic was as controversial in the past as it is in the present (Krippner and Friedman 2010). Many nineteenth-century writers explained claims for these phenomena using such conventional explanations as fraud, coincidences, hallucinations, illusions, suggestion, and a variety of psychophysiological processes related to hysteria and hypnosis (e.g., Carpenter 1877; Dendy 1841; Hall 1887; Janet 1889). There were many attempts to reduce the phenomena in terms of pathology, a topic explored by some in reference to mediumship (for overviews see Alvarado 2010; Alvarado et al. 2007; Le Maléfan 1999; Moreira-Almeida et al. 2005).

Regardless of such general skepticism, interested readers should realize that the topic we are discussing follows trends seen in the historiography of psychology and psychiatry. Starting with Ellenberger (1970), there have been several studies that have made a good case for the idea that psychic phenomena and their study have been a significant factor in the development of ideas about the mind. Concepts such as dissociation and the subconscious mind have been supported in the past by reference to such phenomena as the trances and automatisms of mediums (Binet 1892; Janet 1889; for historical studies see Alvarado 2010; Crabtree 1993; Gauld 1992; Plas 2000; Shamdasani 1993). Psychical research represented more than interest in the supernormal. Many of its researchers also touched on the question about the nature of the mind, its layers, and their separations (Alvarado 2002; Gauld 1992; Plas 2000). The authors of this recent work have changed previous outlooks of psychic phenomena and of movements such as psychical research. Instead of seeing these topics as mere superstitions, or as obstacles in the development of ideas about the mind, this new scholarship places psychical research and the like as catalysts and as important contributing factors to empirical studies that significantly affected the fields of psychology and psychiatry.

In the rest of this chapter, I will summarize some examples of the use of psychic phenomena from the past literatures of mesmerism, spiritualism, and parapsychology to defend the idea of an independent mind. I will consider topics such as ideas to explain apparitions of the living and out-of-body experiences, as well as past discussions of telepathy and mediumship. My point is not to defend the separate existence of the mind from the body. Instead I hope to show the existence of a literature neglected in historical discussions of the mind–body problem linking psychic phenomena to the idea that the mind (and the spirit) is separate from the body. This literature is varied in terms of the arguments presented. Some ideas are derived from actual empirical work, while others are less empirically grounded. But in all cases they are based on observations of phenomena which interpretation has led the writers I will cite to believe that materialistic assumption are insufficient explanations.

3.3 Mesmerism and the Nonphysical Mind

Mesmerism was one of the first large-scale movements to bring psychic phenomena to the attention of the academic Western world. Starting in the eighteenth and further developing in the nineteenth century, mesmerism produced much literature related to psychic phenomena (for overviews see Crabtree 1993; Dingwall 1967–1968; Gauld 1992; Méheust 1999).

Named after Franz Anton Mesmer (1734–1815) (1779/1980), who catalyzed these ideas in the eighteenth century, mesmerism was a movement based on the idea of a universal force called animal magnetism connected to the human body, and responsible for many manifestations, among them magnetic somnambulism (trance) and healing. Although there were many different conceptualizations of this force – a principle generally rejected by current students of hypnosis – most writers identified it with the nervous fluid or animal electricity believed by many to be behind the functioning of the nervous system (e.g., Charpignon 1848; Esdaile 1852).

Although animal magnetism was seen by many to be a physical principle, there were representatives of the mesmeric movement who defended the existence of nonphysical aspects of human beings. They supported their belief by pointing toward the phenomena showed by the mesmerized subjects which included instances of clairvoyance, knowing the thought of others, and medical diagnosis, among other manifestations.

For example, in his book *Instruction pratique sur le magnétisme animal* (1825), J.P.F. Deleuze (1753–1835), a well-known defender of animal magnetism, said that the phenomena of magnetic somnambulism "offer direct proof of the spirituality of the soul" (p. 99). He defined somnambulism as a state that seemed like sleep and in which the person could talk to the magnetizer. But when the person came back to their usual state they "did not keep any recollection of what had taken place" (p. 98). The belief in the spiritual nature of the condition was based on Deleuze's interpretation of his observations. He saw somnambulism as showing the "distinction of two substances, the double existence of the interior man and of the exterior man in a single individual…" (p. 99).

Lawyer J.C. Colquhoun (1785–1854) argued in *Isis Revelata* (1844) that mesmeric phenomena such as acquisition of knowledge not previously known by the mesmerized individual supported the existence of the soul. In his view

> in the phenomena manifested in the higher degrees of Animal Magnetism, we may find a complete practical refutation of all the material theories of the human mind, a most distinct, cogent, and impressive proof of the independent existence of the soul of man, and, consequently, the strongest philosophical grounds for presuming its immortality; since it has now been demonstrated beyond the possibility of rational doubt, that, in its manifestations, it is not necessarily chained down to any particular part of the sensible and mortal body; but that it is capable of exercising its various functions, in peculiar circumstances, without the assistance or cooperation of any of those material organs, by means of which it usually maintains a correspondence with the external world (Vol. 2, pp. 165–166).

Many others related animal magnetism to the soul (e.g., Ashburner 1867; Haddock 1851). Several argued that magnetism produced a state (magnetic somnambulism) that in turn liberated the faculties of the soul, usually obscured in normal life. For example, travelling magnetizer Charles Lafontaine (1803–1892) argued that when the body was rendered inert through the application of magnetic somnambulism the "life of the body is annihilated, the soul ... separates from the common life to live its own life. Its faculties, all immaterial, appear all the more brighter when the annihilation of matter is more complete" (Lafontaine 1852, p. 62). Sight without the body, Lafontaine believed, was possible because the soul was separated from the body.

Consistent with this, French physician L.J.J. Charpignon (1815–1886) stated in his book *Physiologie, médicine et métaphysique du magnétisme* (1848) that somnambulistic lucidity was "inherent to the nature of the soul" (p. 107).

Referring to the facts of mesmerism one author commented on the issue of survival of death. In fact, he seemed to suggest that the powers mesmerism allowed us to see were glimpses of a future spiritual life, assuming, in turn, a spiritual component in human beings during life: "Man is shown by these to be capable of increased sensitive power. *Cui bono* – to what end, if hereafter this increase of faculty become not permanent? Would it be consistent with the goodness of Providence to tantalise us by imperfect glimpses of that which we shall never be permitted to realise? Would wings be folded in the worm if they were not one day to enable it to fly? We cannot think so poorly of creative wisdom or of thrifty nature" (Townsend 1840, pp. 533–534).

German physician and author Johann Heinrich Jung-Stilling (1740–1817) was of the opinion that: "Animal magnetism undeniably proves that we have an inward man, a soul, which is constituted of the divine spark, the immortal spirit, possessing reason and will, and of a luminous body, which is inseparable from it" (Jung-Stilling 1834/1808, p. 227). Some, as did French magnetiser Jules Dupotet de Sennevoy (1796–1881), believed that the teachings of animal magnetism could bring man closer to God, a point he made in his book *Essai sur l'ensegnement philosophique du magnétisme* (Dupotet de Sennevoy and Baron 1845). Such ideas were not scientific but were based on observations of phenomena that the mesmerists believed were unexplained by established knowledge.

3.4   Wandering Spirits and Souls

"The highest species of apparitions ...," wrote Jung-Stilling (1834/1808, p. 73), "is, incontestibly, when a person still living can show himself in some distant place." Apparitions of the living were frequently used to support the idea that the spirit, soul, or some conscious aspect of human beings was able to function out-of-the body (for a discussion of selected aspects of this literature see Alvarado 2009). These were cases in which an apparition, usually a visual representation of an individual, was perceived by someone when the person was not physically present in the location.

Scottish-born known socialist and social reformer Robert Dale Owen (1801–1877) discussed these apparitions in his widely read book *Footfalls on the Boundary of Another World* (1860). In Owen's view such apparitions showed that the

> spiritual body ... may, during life, occasionally detach itself, to some extent or other and for a time, from the material flesh and blood which for a few years it pervades in intimate association; and if death be but the issuing forth of the spiritual body from its temporary associate; then, at the moment of its exit, it is that spirit body which through life may have been occasionally and partially detached from the natural body, and which at last is thus entirely and forever divorced from it, that passes into another state of existence (pp. 360–361).

In many of these cases the appearer was dying or passing through some sort of crisis such as illness or accident. The most systematic investigation of these cases was that done by SPR members, as seen in the first large-scale work of the Society, *Phantasms of the Living* (Gurney et al. 1886; for earlier, and less well investigated cases see Harrison 1879). They collected hundreds of cases of "veridical" experiences. These were cases in which the perceiver learned that the person was having a crisis around the time of the experience, or obtained some information he or she did not have before such as a sense of location or what the appearer looked like at the time of the crisis. The researchers not only interviewed the person that had the experience, getting first hand accounts, but also other persons that could corroborate aspects of the experience or that could bear witness that they had heard the account of the apparitional experience before it was determined that it was "veridical." An example of such an experience is the following:

> In 1856 I was engaged on duty at a place called Roha, some 40 miles south of Bombay .... My sister and I were regular correspondents, and the post generally arrived about 6 a.m. .... It was on the 18th April of that year ... that I received a letter from my mother, stating that my sister was not feeling well, but hoped to write to me the next day .... At 2 o'clock my clerk was with me, reading some native documents that required my attention, and I was in no way thinking of my sister, when all of a sudden I was startled by seeing my sister (as it appeared) walk in front of me from one door of the tent to the other, dressed in her nightdress. The apparition had such an effect upon me that I felt persuaded that my sister had died at that time. I wrote at once to my father, stating what I had seen, and in due time I also heard from him that my sister had died at that time. (Gurney et al. 1886, Vol. 1, pp. 41–42).

While some SPR researchers – such as Edmund Gurney (1847–1888) and Frank Podmore (1856–1910) – argued that these apparitions represented the externalization of a telepathically acquired message by the percipient (Gurney et al. 1886; Podmore 1894), others assumed the existence of a spirit capable of travelling beyond the body carrying consciousness. This assumption clearly assumed the independence of the thinking part of human beings from the workings of the physical body.

Many other later writers had similar ideas, but with some variants. This was the case of French engineer Gabriel Delanne (1857–1926) and Italian student of psychic phenomena Ernesto Bozzano (1862–1943). They presented many published cases arguing that they supported the idea that apparitions of the living were more than imagination because they were veridical (Bozzano 1937/1934; Delanne 1909).

Based on carefully investigated cases by SPR researchers, Myers (1903) speculated that different types of apparitions of the living, such as those that occurred spontaneously while the person was dying or was thinking of arriving to a place were manifestations of the spirit with different degrees of consciousness. He believed that the place where the person was seen was a modification of a "certain portion of space, not materially nor optically, but in such a matter that specially susceptible persons may perceive it" (Myers 1903, Vol. 1, pp. xix–xx). In his view death was a permanent "self-projection," or "the one definite act which it seems as though a man might perform equally well before and after bodily death" (Myers 1903, Vol. 1, p. 297).

In more recent times out-of-body experiences have been discussed as phenomena suggesting that consciousness can function out of the physical body (e.g., Woodhouse 1994). Such idea has been supported with evidence of veridical experiences such as those in which information has been acquired while the person was out of their body, particularly information referring to distant events (Hart 1954). The topic has been studied in the laboratory using designs in which a person is asked to go while having an out-of-body experience to a particular place to perceive information placed there (Alvarado 1982). Although only a handful of studies have shown positive results, this area deserves further research.

Conceptualizing individuals as having a physiological (B system) and a mental life (M/L system) Tart (1979) referred to out-of-body experiences as "a temporary spatial/functional separation of the M/L system from the B system" (p. 195).

Another relevant line of study is investigations of near-death experiences. Several studies have shown the phenomenon has stable features (e.g., Ring 1979; for an overview see Zingrone and Alvarado 2009). Based on his study of this phenomena Dutch cardiologist Pim Van Lommel (2009) has speculated that consciousness is "stored in a non-local space as wave fields of information" (p. 183). In the United States psychiatrist Bruce Greyson referred to the implications of the experience when he stated that they challenge material reductionism "in asking how complex consciousness, including mentation, sensory perception, and memory, can occur under conditions in which current physiological models of mind deem it impossible. This conflict between a materialist model of brain–mind identity and the occurrence of NDEs under conditions of general anesthesia or cardiac arrest is profound and inescapable" (Greyson 2010, p. 43).

3.5   Telepathy and Clairvoyance as Spiritual Faculties

English writer Catherine Crowe (1790–1872) stated in *The Night-Side of Nature* (1848) that the acquisition of information such as clairvoyance had a spiritual component. Some states, she argued citing previous authors, suggested the functioning of spiritual faculties, or "perceptions which are not comprised within the functions of our bodily organs" (Vol. 1, p. 32). Referring to the acquisition of information in dreams, such as cases about dreams of the future, Crowe wrote that when the senses were in a passive state "the universal sense of the immortal spirit within, which sees, and hears, and knows, or rather, in one word, *perceives,* without organs, becomes more or less free to work unclogged" (Vol. 1, p. 98).

Many authors, such as English physician Joseph Haddock (1800–1861), explained psychic phenomena as part of the interaction of the physical and spiritual nature of human beings, but coming from the latter. Haddock (1851) conducted studies of clairvoyance with a woman he mesmerized, obtaining results that led him to consider the idea of a nonphysical mind. He wrote:

> All those apparently miraculous powers, which we sometimes see, or hear of being displayed by good mesmeric subjects, are, in fact, but the result of the psyché, or animus, being so far set free from the bodily ultimate, as to enable the spiritual body to act nearly, if not quite, independently of the sensual organs, and by perception, and in light from an inner world; but the connection of the mind and body is yet sufficient, to enable the soul's sight and feeling to be manifested to our physical senses, by and through the natural organization of a clairvoyant subject (Haddock 1851, p. 72).

Others, such as French educator Hippolyte-Léon Denizard Rivail (1804–1869), known as Allan Kardec (1857), summarized communications claimed to originate from discarnate spirits, and discussed what he referred to as the "emancipation of the soul." This could take place, he wrote, in dreams and in other states. In his view second sight was an example of this emancipation, and took place mainly in "times of crises, calamities, great emotions, finally, all the causes that overexcite the mental …" (p. 74). Kardec wrote that clairvoyance in magnetic and natural somnambulism had the same cause, both were a basic characteristic of the soul, "an inherent faculty to all the parts of the incorporeal self that is in us . . ." (p. 71).

In his study of psychic experiencers astronomer Camille Flammarion (1842–1925) concluded: "The soul, by its interior vision, may see *not only what is passing at a great distance,* but it may also know in advance *what is to happen in the future.* The future exists potentially, determined by causes which bring to pass successive events" (Flammarion 1900, p. 481). He asked, "because the soul acts at a distance by some power that belongs to it, are we authorized to conclude that it exists as something real, and that it is not the result of functions of the brain?" (p. 481). In his view: "These phenomena prove, I think, that the soul exists, and that it is endowed with faculties at present unknown. …A thought can be transmitted to the mind of another. There are *mental transmissions, communications of thoughts, and psychic currents* between human souls" (p. 485).

Years later William MacDougall (1911) argued that telepathy was incompatible with mechanistic ideas. He believed, in the case of veridical automatic messages received by some mediums studied by the SPR, that physical ideas such as brain waves could not account for the coordinated nature of messages received by geographically separated individuals. McDougall did not specify further his objection to a physicalistic explanation of telepathy. But Myers referred to many cases of spontaneous telepathy that did not seem "diminished by any distance nor to be impeded by any obstacle whatsoever" (Gurney et al. 1886, Vol. 1, p. L). He saw telepathy as a spiritual faculty, an indication of a transcendental spiritual realm that worked in human beings subconsciously and that represented a faculty humans would employ in the other world (Myers 1903).

During the twentieth century, many others saw telepathy and clairvoyance as evidence of a nonphysical principle (e.g., Beloff 1990; Bozzano 1942; Pratt 1967). An important representative of this was the well-known American parapsychologist J.B. Rhine (1895–1980). In his book *The Reach of the Mind*, Rhine (1947) not only described his experimental work with ESP (and psychokinesis), but also argued that the results of that research showed the phenomena he studied was the result of a nonphysical principle separate from but interacting with the body. In his view ESP research showed that the "mind can escape physical boundaries under certain conditions .... Accordingly a distinct difference between mind and matter, a relative dualism, has been demonstrated ..." (p. 205). He believed this because in his experiments ESP had not shown any relationship to such physical variables as distance and time, the latter being a reference to precognition. In fact Rhine (1954) referred to parapsychology as the science of nonphysical nature, a discipline concerned with phenomena "that fail to show regular relationships with time, space, mass, and other criteria of physicality" (p. 801). In his view the presence of such a nonphysical component in man validated aspects of religion and could be used to combat communism and materialism in general (Rhine 1953).

In addition to the belief that telepathy and clairvoyance indicated the existence of a spiritual component in human beings, some argued that such phenomena were spiritual faculties that belonged to life in a spiritual realm to be returned to after death (see also my previous reference to Townsend in the section about mesmerism). They showed but occasional glimpses in our earthly life. For example, in *The Philosophy of Mysticism* German philosopher Carl du Prel (1839–1899) stated that "in man himself there is a kernel, to which the laws of sensibility do not apply" (Du Prel 1889/1885, Vol. 2, p. 284) and which is suppressed in life but released at death. The manifestations in question suggested to him that "in human nature there lie already veiled indications of the next higher state of being ..." (Vol. 2, p. 285). Similarly, Myers (1903) referred to the idea that "that portion of the personality which exercises these powers during our earthly existence does actually continue to exercise them after our bodily decay" (Vol. 1, p. 222).

3.6   Survival of Death and Mediumship

In a book about psychical research, Carrington (1908) wrote about the importance of the idea of survival of bodily death to combat materialism. In his words: "If it can be shown that consciousness can persist apart from brain activity and a nervous system, then materialism will have been overthrown and another interpretation of the universe rendered possible" (pp. 10–11). Many past (and some current) efforts of researchers in psychical research have been directed to empirical studies on the topic. Different from purely religious and philosophical speculations, psychical researchers have obtained empirical evidence for phenomena such as apparitions and mediumship that suggest to some that the mind or consciousness can function separate from the body during life and can continue after death (for an overview see Braude 2003).

In actuality the issue of survival of death has been central to psychical research from the beginnings of the discipline (Alvarado 2003). One of the main phenomena in this regard has been mediumship, or the idea that certain individuals can produce messages from the dead. The old psychical research literature, such as some studies of American medium Leonora E. Piper (1857–1950) presents examples of séance studies in which spirit survival has been taken seriously by psychical researchers known for their rigorous empirical approach to the topic (Hodgson 1898; Hyslop 1901; Lodge 1890).

An example of a positive evaluation of the results of sittings with Mrs. Piper and other mediums was offered by the above-mentioned English physicist Sir Oliver J. Lodge, F.R.S. in his book *The Survival of Man* (1920). Lodge wrote:

> The first thing we learn, perhaps the only thing we clearly learn in the first instance, is continuity. There is no such sudden break in the conditions of existence as may have been anticipated; and no break at all in the continuous and conscious identity of genuine character and personality. Essential belongings, such as memory, culture, education, habits, character, and affection, – all these, and to a certain extent tastes and interests, – for better for worse are retained. Terrestrial accretions, such as worldly possessions, bodily pain and disabilities, these for the most part naturally drop away.
>
> Meanwhile it would appear that knowledge is not suddenly advanced … we are not suddenly flooded with new information, – nor do we at all change our identity; but powers and faculties are enlarged, and the scope of our outlook on the universe may be widened and deepened, if effort here has rendered the acquisition of such extra insight legitimate and possible.
>
> On the other hand, there are doubtless some whom the removal of temporary accretion and accidents of existence will leave in a feeble and impoverished condition; for the things are gone in which they trusted, and they are left poor indeed … (pp. 342–343)

Many studies with mediums believed to have provided evidence for survival of death appeared in the twentieth century. Some examples are mediums such as Gladys Osborne Leonard (1882–1968) (e.g., Radclyffe-Hall and Troubridge 1919) and studies of more recent mediums (Beischel and Schwartz 2007).

Many other lines of research have been conducted in which phenomena suggestive of survival of death were studied empirically. For example, among them are cases of children that claim to remember previous lives, and the study of apparitions of the dead (for a review see Braude 2003).

3.7   The Ensemble of Psychic Manifestations

In addition to the above, some students of psychic phenomena took a more comprehensive approach. They believed that a nonphysical view was supported by the whole mass of phenomena taken together, as opposed to analyses of single phenomena. As Crowe wrote: "The subjects I do intend to treat of are the various kinds of prophetic dreams, presentiments, second-sight, and apparitions; and, in short, all that class of phenomena, which appears to throw some light on our physical nature, and on the probable state of the soul after death" (Crowe 1848, Vol. 1, p. 22). Such phenomena indicated to her the existence of a "dweller in the temple," or the idea that "we are immortal spirits, incorporated for a season in a material body" (Vol. 1, p. 32), which is capable of producing the above-mentioned phenomena when it is still in the body and when it is out of it.

Myers used a general approach to argue for the existence of a spirit and for survival of death (Du Prel 1889/1885; Myers 1903). Instead of focusing solely on specific phenomena they considered a wide range of manifestations to make their case. In doing this they did not limit their discussion to psychic phenomena such as telepathy, but considered other psychological manifestations as well, among them hypnosis, hallucinations, and dreams.

In fact, referring to Myers it has been stated that: "There is a continuous set of gradations between facts which everyone accepts and facts which might be called 'paranormal'; and it is hard indeed to find a 'logical halting-place' anywhere along the line" (Gauld 1968, p. 277). William James praised Myers' discussion of many phenomena, and his use of the idea of the subliminal mind, his version of the subconscious that was concerned with psychic and psychological phenomena, but also with spiritual and evolutive faculties. He wrote:

> Myers wove such an extraordinarily detached and discontinuous series of phenomena together. Unconscious cerebration, dreams, hypnotism, hysteria, inspirations of genius, the willing game, planchette, crystal-gazing, hallucinatory voices, apparitions of the dying, medium-trances, demoniacal possession, clairvoyance, thoughts transference – even ghosts and other facts more doubtful – these things form a chaos at first sight most discouraging. No wonder that scientists can think of no other principle of unity among them than their common appeal to men's perverse propensity to superstition. Yet Myers has actually made a system of them, stringing them continuously upon a perfectly legitimate objective hypothesis, verified in some cases and extended to others by analogy (James 1901, p. 18).

This general approach was also used by Ernesto Bozzano, who argued for the importance of "keeping constantly in mind all the data of the problem to be solved which in our case consist of innumerable varieties of metapsychic episodes inexplicable by any naturalistic hypothesis" (Bozzano, n.d., p. 261). In his view if psychic phenomena were studied as an ensemble they were "transformed into a cumulative and logically irresistible poof of the experimentally certified intervention of the spirits of the dead in supernormal manifestations" (p. 261). This involved the assumption that all the characteristics and modalities of manifestation of the phenomena pointed toward a spiritual agency.

Following Myers, Kelly et al. (2007) have recently argued that there are many mental and psychophysiological manifestations that cannot be reduced to neurological mechanisms. Among them they discussed veridical mediumship, out-of-body experiences, but also some of the phenomena of hypnosis, and processes such as memory. Once again, "normal" and "paranormal" phenomena were considered together to paint a canvas of the mind in terms of its lack of dependence on the body and its functions.

3.8 Concluding Remarks

In this chapter, I have presented a short overview of past discussions of psychic phenomena to support the existence of the mind, the spirit, or some sort of non-physical principle in human beings. My review has obviously not covered all relevant aspects of the topic. For example, I have not covered writings and research on remote viewing, reincarnation cases, psychokinesis, or the effects of prayer or therapeutic intent at a distance. Furthermore, it must be remembered that dichotomies such as material and immaterial, and physical and nonphysical, are too simple particularly when we bring into consideration ideas from modern physics as discussed, for example, in relation to psychic phenomena (Radin 2006).

The ideas presented in this chapter are speculations to account for unexplained phenomena. As such they are not accepted by every student of the subject. Because these are matters that cannot be measured directly we are left with interpretations about nonphysicality, which lack relevant details about the nature of the process. That is, many of the ideas discussed are vague about the nature of the mind or the spirit beyond the statement that the phenomena do not seem explainable through conventional physical ways. Although this is problematic, and limits conceptual and research progress in this area, in reality it is not much different from many speculations in science. Many areas of science face the study of phenomena that are not understood and it is necessary to start from speculations based on what is observable, a process that also applies to materialistic ideas.

But regardless of these issues, the best efforts of psychical research need to be recognized as a serious and empirical effort to study the properties of the mind, and as a literature that presents facts that are not easily explained via conventional psychological, physical and physiological processes.

Although I believe there is acceptable evidence to support the existence of psychic phenomena beyond conventional explanations such as fraud, coincidence, and hallucination, it has not been my intention in this chapter to provide or discuss this evidence in detail. Instead I have presented an outline of a literature, and a set of arguments that deserve to be recognized as part of intellectual history and of the history of philosophy and psychology, fields relevant to the mind–body problem. Unfortunately, much of what I have mentioned is generally ignored by most of those who cover these topics, and particularly by the practitioners of disciplines related to the mind–body problem. This is unfortunate because the omission of this material from the mainstream historical accounts of the mind–body problem creates an incomplete history of the subject.

The message discussed in this chapter has been expressed by others recently from the point of view of the existence and validity of the phenomena, which are conceived of as manifestations incapable of being explained by the materialistic paradigm (e.g., Kelly et al. 2007; Tart 2009). Charles T. Tart (2009) has reminded us about this in his book *The End of Materialism* in which he postulates that parapsychological phenomena show that physicalistic approaches cannot explain many important aspects of our nature. In his words: "The findings of scientific parapsychology force us to pragmatically accept that minds can do things … that cannot be reduced to physical explanations, given current scientific knowledge or reasonable extensions of it" (p. 241). Others have presented strong defenses of the nonlocal nature of the mind on the basis of the existence of phenomena as such as ESP (e.g., Schwartz 2007).

These, and many other recent authors, have emphasized the existence of a nonphysical mind, continuing the intellectual tradition described in this chapter. English psychologist John Beloff (1920–2006) characterized this tradition as one which "exhibits mind as an efficacious factor in the real world, not just as an idle epiphenomenon, and thereby calls into question the physicalistic position" (Beloff 1994, p. 518).

Acknowledgments I thank Nancy L. Zingrone for useful editorial suggestions for the improvement of this chapter.

References

Alvarado, C. S. (1982). ESP during out-of-body experiences: A review of experimental studies. *Journal of Parapsychology, 46,* 209–230.

Alvarado, C. S. (2002). Dissociation in Britain during the late nineteenth century: The society for psychical research, 1882–1900. *Journal of Trauma and Dissociation, 3,* 9–33.

Alvarado, C. S. (2003). The concept of survival of bodily death and the development of parapsychology. *Journal of the Society for Psychical Research, 67,* 65–95.

Alvarado, C. S. (2009). The spirit in out-of-body experiences: Historical and conceptual notes. In B. Batey (Ed.), *Spirituality, science and the paranormal* (pp. 3–19). Bloomfield: Academy of Spirituality and Paranormal Studies.

Alvarado, C. S. (2010). Classic text No. 84: 'Divisions of personality and spiritism' by Alfred Binet (1896). *History of Psychiatry, 21,* 487–500.

Alvarado, C. S., & Krippner, S. (2010). Nineteenth century pioneers in the study of dissociation: William James and psychical research. *Journal of Consciousness Studies, 17,* 19–43.

Alvarado, C. S., Machado, F. R., Zangari, W., & Zingrone, N. L. (2007). Perspectivas históricas da influência da mediunidade na construção de idéias psicológicas e psiquiátricas. *Revista de Psiquiatia Clínica, 34,* 42–53.

Ashburner, J. (1867). *Notes and studies in the philosophy of animal magnetism and spiritualism.* London: H. Baillière.

Barrett, W. F. (1918). The deeper issues of psychical research. *Contemporary Review, 113,* 169–179.

Beischel, J., & Schwartz, G. E. (2007). Anomalous information reception by research mediums demonstrated using a novel triple-blind protocol. *Explore: The Journal of Science & Healing, 3*, 23–27.

Beloff, J. (1990). *The relentless question: Reflections on the paranormal*. Jefferson: McFarland.

Beloff, J. (1994). The mind-brain problem. *Journal of Scientific Exploration, 8*, 509–522.

Binet, A. (1892). *Les altérations de la personnalité*. Paris: Félix Alcan.

Bozzano, E. (1937). *Les phénomènes de bilocation*. Paris: Jean Meyer. (Original work published 1934).

Bozzano, E. (1942). *Dei fenomeni di telestesia*. Verona: L'Albero.

Bozzano, E. (n.d., ca 1938). *Discarnate influence in human life: A review of the case for spirit intervention*. London: International Institute for Psychical Research/John M. Watkins.

Braude, S. A. (2003). *Immortal remains: The evidence for life after death*. Lanham: Rowman & Littlefield.

Brower, M. B. (2010). *Unruly spirits: The science of psychic phenomena in modern France*. Chicago: University of Illinois Press.

Carpenter, W. B. (1877). *Mesmerism and spiritualism, &c. historically and scientifically considered*. London: Longmans and Green.

Carrington, H. (1905). The origin and nature of consciousness. *Metaphysical Magazine, 18*, 42–55.

Carrington, H. (1908). *The coming science*. Boston: Small, Maynard.

Charpignon, J. (1848). *Physiologie, médicine et métaphysique du magnétisme*. Paris: Germer Baillière.

Colquhoun, J. C. (1844). *Isis revelata: an inquiry into the origin, progress and present state of animal magnetism* (3rd ed., 2 Vols.). Edinburgh: Maclachlan & Stewart.

Crabtree, A. (1993). *From Mesmer to Freud: Magnetic sleep and the roots of psychological healing*. New Haven: Yale University Press.

Crane, T., & Patterson, S. (Eds.). (2000). *History of the mind-body problem*. London: Routledge.

Crowe, C. (1848). *The night-side of nature, or ghosts and ghost seers* (2 vols.). London: T.C. Newby.

Dupotet de Sennevoy, Baron J.. (1845). *Essai sur l'ensegnement philosophique du magnétisme*. Paris: A. René.

Delanne, G. (1909). *Les apparitions matérialisées des vivants & des morts: Vol. 1. Les fantômes de vivants*. Paris: Librairie Spirite.

Deleuze, J. P. F. (1825). *Instruction pratique sur le magnetisme animal*. Paris: A. Belin.

Dendy, W. C. (1841). *The philosophy of mystery*. London: Longmans, Orme, Brown, Green & Longmans.

Dingwall, E. J. (Ed.). (1967–1968). *Abnormal hypnotic phenomena: a survey of nineteenth century cases*. New York: Barnes and Noble.

Du Prel, C. (1889). *The philosophy of mysticism* (2 Vols.). London: George Redway. (Original work published 1885).

Ellenberger, H. F. (1970). *The discovery of the unconscious: The history and evolution of dynamic psychiatry*. New York: Basic Books.

Esdaile, J. (1852). *Natural and mesmeric clairvoyance*. London: Hippolyte Baillière.

Figuier, L. (1860). *Histoire du merveilleux dans des temps moderns* (4 Vols.). Paris: L. Hachette.

Flammarion, C. (1900). *L'Inconnu: The unknown*. New York: Harper & Brothers.

Gauld, A. (1968). *The founders of psychical research*. London: Routledge & Kegan Paul.

Gauld, A. (1992). *A history of hypnotism*. Cambridge: Cambridge University Press.

Goodrick-Clarke, N. (2008). *The Western esoteric traditions: A historical introduction*. New York: Oxford University Press.

Greyson, B. (2010). Implications of near-death experiences for a postmaterialist psychology. *Psychology of Religion and Spirituality, 2*, 37–45.

Gurney, E., Myers, F. W. H., & Podmore, F. (1886). *Phantasms of the living* (2 vols.). London: Trübner.

Haddock, J. W. (1851). *Somnolism & psycheism* (2nd ed.). London: James S. Hodson.

Hall, G. S. (1887). Review of proceedings of the society for psychical research, and phantasms of the living. In E. Gurney, F.W.H. Myers, & F. Podmore. (Eds.), *American Journal of Psychology* (1), 128–146.

Harrison, W. H. (1879). *Spirits before our eyes*. London: W.H. Harrison.

Hart, H. (1954). ESP projection: Spontaneous cases and the experimental method. *Journal of the American Society for Psychical Research, 48,* 121–146.

Hodgson, R. (1898). A further record of observations of certain phenomena of trance. *Proceedings of the Society for Psychical Research, 13,* 284–582.

Hyslop, J. H. (1901). A further record of observations of certain phenomena of trance. *Proceedings of the Society for Psychical Research, 16,* 1–649.

Inglis, B. (1992). *Natural and supernatural: a history of the paranormal from earliest times to 1914* (ed. rev.). Dorset: Prism.

James, W. (1890). A record of observations of certain phenomena of trance (5) Part III. *Proceedings of the Society for Psychical Research, 6,* 651–659.

James, W. (1898). *Human immortality*. Boston: Houghton Mifflin.

James, W. (1901). Frederic Myers's service to psychology. *Proceedings of the Society for Psychical Research, 17,* 13–23.

James, W. (1909). Report on Mrs Piper's hodgson-control. *Proceedings of the Society for Psychical Research, 23,* 2–121.

Janet, P. (1889). *L'automatisme psychologique: Essai de psychologie expérimentale sur les formes inférieures de l'activité humaine*. Paris: Félix Alcan.

Jung-Stilling, J. H. (1834). *Theory of pneumatology, in reply to the question, what ought to be believed or disbelieved concerning presentiments, visions, and Apparitions*. New York: J. S. Redfield (Original work published in 1808).

Kardec, A. (1857). *Le livre des esprits*. Paris: E. Dentu.

Kelly, E. F., Kelly, E. W., Crabtree, A., Gauld, A., Grosso, M., & Greyson, B. (2007). *Irreducible mind: Toward a psychology for the 21st century*. Lanham: Rowman & Littlefield.

Krippner, S., & Friedman, H. L. (Eds.). (2010). *Debating psychic experience: Human potential or human illusion?* Santa Barbara: Praeger.

Lafontaine, C. (1852). *L'art de magnétiser* (2nd ed.). Paris: Germer Baillière.

Le Maléfan, P. (1999). *Folie et spiritisme: Histoire du discourse psychopathologique sur la pratique du spiritisme, ses abords et ses avatars (1850–1950)*. Paris: L'Hartmattan.

Lodge, O. (1890). A record of observations of certain phenomena of trance (2) Part I. *Proceedings of the Society for Psychical Research, 6,* 443–557.

Lodge, O. (1920). *The survival of man: A study of unrecognized human faculty (New* (enlargedth ed.). New York: George H. Doran.

MacDonald, P. S. (2003). *History of the concept of mind: Speculations about soul, mind and spirit from Homer to Hume*. Aldershot, Hants, England: Ashgate.

McDougall, W. (1911). *Body and mind: A history and a defense of animism*. New York: Macmillan.

Méheust, B. (1999). Somnambulisme et mediumnité (1784–1930): Vol. 1 Le défi du magnétisme animal. Le Plessis-Robinson: Institut Synthélabo pour de Progrès de la Connaissance.

Mesmer, F.A. (1980). Dissertation on the discovery of animal magnetism. In G. Bloch (Ed.), *Mesmerism: a translation of the original Scientific and medical writings of F. A. Mesmer* (pp. 43–78). Los Altos: William Kaufmann. (Original work published 1779).

Moreira-Almeida, A., Almeida, A. A. S., & Lotufo Neto, F. (2005). History of 'spiritist madness' in Brazil. *History of Psychiatry, 16,* 5–25.

Myers, F. W. H. (1900). Presidential address. *Proceedings of the Society for Psychical Research, 15,* 110–127.

Myers, F.W.H. (1903). *Human personality and its survival of bodily death* (2 Vols.). London: Longmans, Green.

Objects of the Society. (1882). *Proceedings of the Society for Psychical Research, 1,* 3–6.

Owen, R. D. (1860). *Footfalls on the boundary of another world: With narrative illustrations.* Philadelphia: J.B. Lippincott.

Perty, M. (1861). *Die mystischen Erscheinungen der menschlichen Natur.* Leipzig: C.F. Winter.

Plas, R. (2000). *Naissance d'une science humaine: La psychologie: Les psychologues et le "merveilleux psychique.* Rennes: Presses Universitaires de Rennes.

Podmore, F. (1894). *Apparitions and thought-transference.* London: Walter Scott.

Podmore, F. (1902). *Modern spiritualism: a history and criticism* (2 Vols.). London: Methuen.

Pratt, J. G. (1967). *Parapsychology: An insider's view of ESP.* New York: E.P. Dutton.

Radclyffe-Hall, M., & Troubridge, U. (1919). On a series of sittings with Mrs. Osborne Leonard. *Proceedings of the Society for Psychical Research, 30,* 339–554.

Radin, D. (2006). *Entangled minds: Extrasensory experiences in a quantum reality.* New York: Paraview Pocket Books.

Rhine, J. B. (1947). *The reach of the mind.* New York: William Sloane.

Rhine, J. B. (1953). *New world of the mind.* New York: William Sloane.

Rhine, J. B. (1954). The science of nonphysical nature. *Journal of Philosophy, 51,* 801–810.

Ring, K. (1979). *Life at death: A scientific investigation of the near-death experience.* New York: Coward, McCann & Geoghegan.

Schwartz, S. (2007). *Opening to the infinite: The art and science of nonlocal awareness.* Buda, TX: Nemoseen Media.

Shamdasani, S. (1993). Automatic writing and the discovery of the unconscious. *Spring, 54,* 100–131.

Tart, C. T. (1979). An emergent-interactionist understanding of human consciousness. In B. Shapin & L. Coly (Eds.), *Brain/mind and parapsychology* (pp. 177–200). New York: Parapsychology Foundation.

Tart, C. T. (2009). *The end of materialism: How evidence of the paranormal is bringing science and spirit together.* Oakland: New Harbinger.

Taylor, E. (1996). *William James on consciousness beyond the margin.* Princeton: Princeton University Press.

Townsend, C. H. (1840). *Facts in mesmerism, with reasons for a dispassionate inquiry into it.* London: Longman, Orme, Brown, Green & Longmans.

Van Lommel, P. (2009). Endless consciousness: a concept based on scientific studies on near-death experiences. In C. D. Murray (Ed.), *Psychological scientific perspectives on out-of-body and near-death experiences* (pp. 171–186). New York: Nova Science.

Watkins, C. S. (2007). *History and the supernatural in medieval England.* New York: Cambridge University Press.

Wolffram, H. (2009). *The stepchildren of science: Psychical research and parapsychology in Germany, c. 1870–1939.* Amsterdam: Rodopi.

Woodhouse, M. (1994). Out-of-body experiences and the mind-body problem. *New Ideas in Psychology, 12,* 1–17.

Wright, J. P., & Potter, P. (Eds.). (2000). *Psyche and soma: Physicians and metaphysicians on the mind-body problem from antiquity to Enlightenment.* New York: Oxford University Press.

Zingrone, N. L., & Alvarado, C. S. (2009). Pleasurable Western adult near-death experiences: Features, circumstances and incidence. In J. M. Holden, B. Greyson, & D. James (Eds.), *The handbook of near-death experiences* (pp. 17–40). Santa Barbara: Praeger.

Part II
Physics

Chapter 4
No-Collapse Physics and Consciousness

Chris Clarke

Abstract This chapter first reviews strands of work in philosophy, psychology and physics which, taken together, undermine the world-views of both the Cartesian duality of body and soul, and the mechanistic picture of a system governed only by universal deterministic law. Starting from an alternative polarity suggested by recent work in psychology and using established ideas from modern quantum physics, a provisional structure is presented for an alternative world view. The consequences of this for future research are indicated.

4.1 Background

Since this chapter draws on work from several different disciplines relating to physics and consciousness studies, I will begin by summarizing this background.

4.1.1 The Philosophical Context

The reference to "the Mind-Brain Relationship" in the title of this book demands engagement with philosophy: the title implicitly proposes the thesis that mind and brain are distinct, and whether this is denied or supported, whatever the discipline from which evidence or arguments are brought, philosophical issues are implicated. Starting from the reference-point of Cartesian dualism, I shall be drawing in particular on those strands which, sometimes indirectly, suggest some form of duality in the universe, and on those which negatively critique the particular duality espoused by Descartes.

C. Clarke (✉)
School of Mathematics, University of Southampton, Southampton, SO17 1BJ, UK
e-mail: cclarke@scispirit.com

A. Moreira-Almeida and F.S. Santos (eds.), *Exploring Frontiers of the Mind-Brain Relationship*, Mindfulness in Behavioral Health, DOI 10.1007/978-1-4614-0647-1_4, © Springer Science+Business Media, LLC 2012

Descartes gave what was to be the definitive form to the idea that the world is based on exactly two substances, by definition independent of anything else (other than God) for their existence, which Descartes termed *res cogitans* and *res extensa*, designating his conceptions of mind and matter, respectively. *Res extensa*, could manifest both as corpuscles and as aether (similar to Newton's corresponding concept composed of both corpuscles and the metaphysically ambiguous "space"). *Res cogitans*, on the other hand, manifested as the soul. Thus, in particular, matter (*res extensa*) was the foundation for the brain and *res cogitans* was the foundation for mind. Mind and brain were distinct by virtue of a distinction of substances. While the basis for this was the widely accepted view of Aristotle and the mediaeval Aristotelian schools, Descartes' achievement was to propose a detailed mechanical process which enabled this duality to work. He greatly enlarged the range of human sensations and actions that could be understood in purely mechanical terms by explaining how they could be executed by the body, while those which he regarded as mechanically impossible, such as the recognition of speech, were left to the soul. In this he sowed the seeds of the subsequent withering away of the soul in subsequent thinking, as more and more of its supposed functions were transferred to the body, giving progressively more support to those (such as Hobbes) who held a purely mechanistic position, one now probably adopted by the majority of scientists.

Despite this, there are significant grounds for approaching mechanism critically. The erosion of the idea of the soul has been described in this volume by Saulo de Freitas Araujo and critiqued extensively by him on the grounds that it fails to explain the basic fact of subjectivity and is postulated on the basis of unsupported assumptions. In addition, Erlendur Haraldsson and Peter Fenwick (see Chaps. 9–11) have reported careful investigations of phenomena that are hard to reconcile with naïve mechanism using current science and physiology, such as a person's acquisition of information at a time when they had no way of achieving this; either because they were physically isolated from the source of the information or because, in the case of near-death experiences, they were clinically dead at the time, or both. But if the mere reduction of Cartesianism to materialism is not supportable, neither is an arbitrary clinging to Descartes' position of a separate body and soul. Unless a great deal more is said about the relationship between a postulated soul or similar entity and the conventionally understood world, the soul remains nothing but a magic box whose only property is to keep intervening in physics in order to make life more interesting. In the middle ages, there were at least serious attempts by Aquinas and others to put together a coherent picture of the universe in which body and soul had carefully articulated roles, a need which Descartes fully grasped.

Another version of "magic box dualism" in an attempt to explain difficult phenomena is to postulate a global "field" (the word here having no connection with the scientific usage, despite claims to the contrary criticized by Clarke and King (2006)), which either stores information independently of a person's physical body or has means of acquiring information other than through the normal processes. Neither this, nor the opposite, equally unscientific reaction of simply dismissing all reports of phenomena that appear not to fit with current theories, will enable our understanding to progress.

I will be presenting here a number of strands that point to a position that is neither Cartesianism nor materialism, but is based on a combination of conventional modern science with a duality that is quite different from, and more experimentally grounded than that of Descartes. The first of these strands is Kant's core proposition, as interpreted by Savile (2005), that the world as it is known by us is structured a priori by our capacities of knowing. Unlike Kant, however, I recognize that these capacities change with time through changes in human culture and genetics. Thus at any given cultural stage of our understanding, we are imposing a particular filter on the nature of the world as we perceive it, and this demands a corresponding form from our explanations of the world. In the next subsection, I shall be presenting arguments from psychology that there is a basic duality within our knowing, which is reflected in our theories.

Another strand, which at first seems in its realism to be incompatible with the quasi-Kantian strand just cited, is the argument that our cognitive abilities to evolve by natural selection so as to be sensitive to the structure of the world, perhaps evolving distinct cognitive structures to sense distinct aspects of the world. Hence the fact that we possess a duality of knowing might be an indication of something about the universe, which can be well handled in a dual cognitive manner. This argument is reconciled with the quasi-Kantian one by the participatory aspects of the work of Varela et al. (1993) and especially Ferrer (2002), who suggest that what we call reality is cocreation between human cognition and existence; that reality lies between our grasping and the ungraspable.

A further strand emerges with the rise of quantum theory in the twentieth century, which saw a softening of the dichotomy between mind and matter. The universe, including humans as inconsequential parts of it, was a somewhat wobbly machine. Its gears turned almost predictably from one moment to the next, disturbed only marginally by moments of random "quantum collapse." These quantum events were supposedly fully explained by Niels Bohr in 1927, but at the cost of restricting physics to operations in a laboratory with no overarching framework for the world as a whole. Moreover, as I discuss further in Sect. 4.1.3 below, a minority of writers, such as London and Bauer (1939, 1983) and Squires (1994), linked consciousness or mind with the quantum formalism in a way that echoed Cartesianism. Mind started to seem relevant, but the relationship between mind and matter was as obscure as it had been in Descartes' approach.

Today the time has come to grasp these several strands and relate them to a new integrated whole. Not only does there need to be the equivalent of a science of the soul (for which some elements are present in existing writings) but in order to constitute an genuine explanation it needs to be meshed with science as we have it now, not science as it was 70 years ago. In this chapter, I will be setting out one way in which this might be done, by the application of current physics to a recently explored form of duality based not on two substances but on two principal ways of knowing. I indicate later (Sect. 4.3, note 4) the way, using this structure, modern quantum theory can illuminate some of the phenomena explored in this volume. In addition many "parapsychological" phenomena can be explained (Clarke 2008) by this approach.

4.1.2   The Background in Psychology and Consciousness Studies

Three specific topics will be drawn on in this area: the idea of different ways of knowing, the developments in understanding consciousness over the past 20 years, and the connection between these and modern systems of logic.

4.1.2.1   The Dual Nature of Knowing

The quasi-Kantian approach described above directs attention to our cognitive structures. Here several theories imply that we have two different primary modes of knowing, imparting a dynamical polarity to the functioning of the human mind, which, following the preceding remarks on Kant, suggests that the world might manifest to us two different aspects. The polarity theories for which there is strong evidence are the "interacting cognitive subsystems" (ICS) model of Teasdale and Barnard (1993) and the more recent cerebral hemisphere model of McGilchrist (2009). The former uses data from experimental psychology to support a scheme that includes and refines several psychotherapeutic models, while the latter is based on a detailed survey of behavioral and neurological evidence that the two hemispheres of the brain induce a polarity of knowing strikingly similar to the ICS model. Each sheds light on the other. In both cases, the authors recognize that it is in fact a gross over-simplification to describe the theories simply as dualisms. Human knowing is a vast and complex thing. What is remarkable is that these models based on comparatively simple idealized operations (Barnard) or the large-scale anatomy of the brain (McGilchrist) can shed such light on fundamental aspects of knowing.

According to Teasdale and Barnard (1993), we are governed, at the top level of our mental organization, by two distinct meaning-making "interacting cognitive subsystems." One (the "implicational" subsystem) is concerned with the significance for the self of its overall context, drawing immediately of our sensations. It deals with what concerns us, including monitoring threats and opportunities, and with relationships, in the sense of our meaningful connections both with other beings and within ourselves. The other subsystem (the "propositional") is concerned with analysing experiential data from the implicational subsystem – abstracting from it general concepts (including the self-image) and linking them into conceptual propositions, which are in turn fed back to the implicational subsystem, thus enabling thinking to be reflective. It is closely connected with speech but has no direct contact with the senses. In terms of logic, the propositional is the basis of rational thought while the implicational is prerational. Each subsystem has its own memory store and there are measurable characteristic time delays for memory access and transfer times between systems.

The model of McGilchrist (2009) is similar, associating the higher level cognitive functions of the left and right hemispheres with properties very close to those of the propositional and implicational subsystems, respectively. He stresses that, in their cognitive functions, the right hemisphere (Barnard's implicational) deals in the presence of whole entities, the left in abstractions and categorizations. It seems that, without too much over-simplification, the two models are different views of the same systems of knowing, whose components I will refer to as rational (left hemisphere-based, propositional) and relational (right hemisphere-based, implicational).

The key theme of *The Master and his Emissary* (McGilchrist 2009) is that the rational system, while in many ways inferior to and subordinate to the relational system, has in the recent course of biological and, more importantly, cultural evolution usurped the role of the relational system – hence his title, framing a metaphor in which the rational is like a ruler's emissary who has abrogated the ruler's power to themselves. This means that, in our thinking (for instance, as I am now thinking as author and as you, dear reader, are thinking as reader) we find it hard to access rationally the suppressed but still structurally vital relational system. We think as if there was no duality and it is hard to accommodate the notion of the relational into our rational thinking: we tend to conceptualize it as a sort of fuzzy version of the rational, rather than a way of knowing in its own right. Concepts like alternative logic can help in translating relational knowing more accurately into rational terms. I discuss this further in Sect. 4.1.2.3.

4.1.2.2   The Place of Consciousness

Consciousness is associated with both subsystems/hemispheres. On Barnard's model (Teasdale and Barnard 1993), "consciousness" is associated with a process of recording impressions in long-term memory, which switches rapidly and continually from one system to the other giving the impression of continuity. It will be helpful to clarify the sense in which I am interpreting "consciousness" here. This word is used in many different ways. Some are applicable only to humans, others to more general biological systems (organisms or parts of organisms). Following Velmans (2000) I am taking "consciousness" to mean the active facility for experiencing in an individual system. This sense of the word does not refer to information processing (which could equally well happen "in the dark" as Velmans puts it) nor to the physical correlates of consciousness, nor to why we are conscious of some things and not others, and it is not restricted to the reflexive consciousness of one's self. It is the bare fact of awareness, not the content of that awareness. While both systems/hemispheres are responsible for consciousness, they exercise their function in different ways. In McGilchrist's account, the right hemisphere responds to the immediate "presence" of the world as it appears to us, in the language of the debate between Steiner and Derrida (Steiner 1989), rather than its conceptual structure. This response is passed to and represented in the left hemisphere, where classification and analysis takes place, and this aspect also is contributed to the total conscious experience.

Over the past 30 years, the study of consciousness has increasingly drawn on other academic areas in addition to psychology, turning it into "consciousness studies," a discipline in its own right. Within this area a decisive strand of thought has emerged drawing on *What is it like to be a bat?* (Nagel 1974) and *Facing up to the problem of consciousness* (Chalmers 1995). This strand takes consciousness in the sense defined by Velmans (2000), who derives it from Nagel (1974), in which consciousness is categorically different from brain function. The former is essentially subjective, the latter objective (directly shared in the public domain).

It appears from this work that questions about the distinction, or lack of it, between mind and brain are more clearly formulated in terms of the distinction between consciousness and brain functions because this distinction seems easier to define. As an example of the use of this distinction, we might consider a given computational operation (function) carried out by the brain, which could be functionally identical to an operation carried out on the silicon chips of a computer. Then we might be inclined to think, from our prejudiced position of not being based on silicon chips ourselves, that only the former could be accompanied by consciousness, while the later would always be, in Velmans' phrase, in the dark. If such an argument could be supported by wider theoretical considerations, it would establish a distinction based on consciousness that captured the spirit of the mind/brain distinction.

In contrast to this approach to mind based on consciousness, consider the situation with Descartes' mind (*res cogitans* or "soul") where the distinction between soul and body was based on what functions had to be exercised by the soul, rather than how they were exercised (as in the case with considerations of consciousness). On this basis, Descartes included in the scope of mind operations like speech recognition. But as the understanding of computation grew, so functions reserved to Descartes' *res cogitans* were transferred to *res extensa* until today there would appear to be nothing left. The advantage of drawing the line between conscious and nonconscious operations, though not without its own problems, is at least not dependent on our ingenuity in thinking of computational algorithms. Consciousness is a qualitative subjective aspect of a physical system.

One problem with drawing the line on the basis of consciousness is revealed by the concept of "the unconscious mind," which often designates a repository of processes which are in all essential respects of the same form as conscious ones, but of which I am not conscious. If we were to define "mind" in terms of consciousness, the phrase "unconscious mind" would then become an oxymoron. To handle this, I suggest that here we need to unpack the "I" in the phrase "processes ... of which I am not conscious." Here "I" refers to a system that embraces speech, typing book chapters and so on, whose functions and memories constitute a well-connected set. It is likely, however, that the human being possesses, in Lockwood's (1989) phrase, a "compound I" including additional conscious systems that are interconnected or overlapping to limited and varying degrees. I will call the dominant, "up front I" that is linked with speech the *explicit consciousness*.

Conscious systems are, on this conception, associated with particular processes in an organism, to at least some extent distinct from each other and associated with distinct brain structures. Rather than Descartes' single "seat of consciousness" (in his case, in the pineal gland) there are multiple loci of consciousness. The work of Hameroff in the next chapter is phrased in terms of the conscious/nonconscious distinction rather than that of "mind" and "brain," and, as in the main part of this chapter, it seeks to locate the elementary loci of consciousness and explore how widely each one (each "I") can extend at the physical level.

Approaching consciousness in this way gives further insight into the polarity of cognitive systems/hemispheres described above. Both the central systems are themselves conscious, and their interconnection seems to be such that our explicit consciousness relates to both systems. A distinction arises, however, when we consider not the consciousness of the system itself, but how the system discerns the presence or absence of consciousness in beings and events in the world – discernments, for example, of agency, creativity, manifestations of emotion, and so on in other beings. Such discernments are often described as linked to a "theory of mind" (Premack and Woodruff 1978) held by the subject of discernment. The relational system, with its capacity for establishing empathic connections with whole entities and their contexts, is far better at appreciating the presence of consciousness and understanding it in the flexible, intuitive way that is appropriate to its nature. The rational system tries only to analyse consciousness verbally, inevitably failing, while being far superior at grasping the nature of the mechanical, computational aspect of the world.

To make matters even more complicated, the characteristic of human consciousness is that it is reflexive: not only do we apply our psychological "theory of mind" to others, but we apply it to ourselves, and we do so in a dual manner, as discussed at the end of Sect. 4.1.2.1. Thus we have our rational consciousness of the relational, our relational consciousness of the relational, and so on.

4.1.2.3 Knowing and Logic

The situation described in the preceding subsection not only gives a physiological clue to where we might find the physical correlates of consciousness, but it also imposes a dramatic methodological requirement on the study of consciousness. If we cannot grasp the nature of consciousness purely rationally, then we will need a nonverbal, nonanalytical investigation to complement a standard scientific investigation of the physical correlates of consciousness. There is, of course, a very extensive literature on the nonverbal approaches (particularly within Buddhism) based on detailed internal nonverbal investigation, more usually referred to as "spirituality." It requires some personal acquaintance with actual practice for its comprehension; given this, it is making a large contribution to this area. This has informed the discussion in this chapter, though I will not be considering it in detail.

If we apply this duality of ways of knowing to the quasi-Kantian position outlined earlier, we are led to envisage the world as possessing a corresponding duality. Rather than matter and spirit, its poles might be described in terms of quality of function as mechanical and ontological, the latter reflecting the sense of presence associated with the relational system. The fundamental stance of current mainstream science is to attempt to either deny the existence of the ontological pole, or to reduce it to the mechanical. From this point of view, this mainstream stance could be called mechanism, rather than materialism.

A major problem to be analysed below is that of relating what we know about consciousness from internal investigation of ourselves to the physics of a "locus of consciousness" – a physical system that is associated with consciousness. In particular, physics as a discipline is carried out using strict classical logic (even though the formal logic within quantum theory itself has the structure of non-standard logic). Moreover, the operations of our own relational system, as we subjectively perceive them, are not governed by logic in the formal sense but, as argued for example by the Chilean analytic psychologist Ignacio Matte Blanco (1998) or by Routley et al. (1982), by a more flexible logic. Some of the features of what Matte Blanco, analysing the unconscious, termed "biologic" are highly relevant to the issue here. It is a context-dependent logic (a feature already implicitly used in quantum theory in the form of topos logic). In common with intuitionistic logics such as Heyting's (van Atten 2009), it does not have a well-behaved operation of negation: if some proposition P is not true, one cannot deduce that the proposition not-P is true. I shall return to these points in Sect. 4.2.1.

4.1.3   The Background in Physics

Contemporary physics is very different from that of Descartes, which raises further problems with the *extensa/cogitans* distinction. To begin with, current physics has blurred the concept of matter, making "corpuscles" a specifically macroscopic concept and installing as fundamental the quantum field (essentially a space-time embodiment of the laws of motion) and the quantum state, a nonlocal function of the quantum field. These concepts have a somewhat noetic aspect in comparison with the mechanical flavor of Cartesian and Newtonian physics. Since these are fundamental to the argument, I will first sketch some basic concepts of quantum physics.

The theory owes its structure and ideas to Niels Bohr, who rooted quantum physics in laboratory practice, stressing the need for connecting quantum physics with the basic quantitative concepts on which classical physics was founded (position, mass, and momentum in particular) and the relations between them. The later formalization of this by von Neumann still forms the basis of the main-stream approach. He defined the core of quantum physics as a process of observation (or measurement) having three stages: the preparation, or selection, of a system in a well defined state; the placing of the system in an observing apparatus, which defines a particular context (what it is that is going to be measured); and the result of the measurement, for which the quantum formalism supplies the probabilities of obtaining the various possible values. These stages are illustrated below, with some of the alternative terminology (corresponding to alternative conceptualizations) added.

If the system is then fed back into a repetition of this process, the probabilities will be different, indicating that the state has been changed by the observation. This is quantum collapse. Alternatively, the system may be left alone to evolve through a dynamic which exactly mirrors the dynamics of Newtonian theory.

Physics today is also different from physics in the earlier twentieth century, when, following the work of London and Bauer (1939, 1983), there was minority support for the idea that collapse was produced by the consciousness of the human experimenter who needed to be brought into the diagram above. This was a dualistic approach. It was supposed that there was a nonphysical process, such as awareness by a "soul" or by "consciousness" as a nonphysical substance, that produces the collapse. The problem here is that this cannot be considered until we have a coherent metaphysics of the soul, including how it can interact with matter without having a physical aspect. In the absence, in my opinion, of such a metaphysics (despite many valiant attempts), this is a "magic box" argument.

The work reported in this chapter and that of Hameroff described in his later chapter is in a way the successor to this, but in a very different form. The older dualistic version of collapse by consciousness was transformed first by the work of Zeh (1970), which showed that an essential requirement, perhaps a defining requirement, for a quantum "observation" was that the observing system was large enough to undergo the process of decoherence, in which interaction between this system and its environment reduced the statistics of the outcome of the observation to a classical form, as opposed to a quantum form. Decoherence depends simply on the size of the observing instrument, not on any consideration of consciousness, which undermined the simple picture of London and Bauer. It did not, however, explain why it is that the system carrying the explicit "I" experiences particular outcomes rather than statistical blurs.

A second difference between current conceptions of quantum observation and earlier ones arises indirectly from work in quantum cosmology. When one regards the whole universe as a quantum object then, theism apart, there is no external observer to which one can appeal in order to explain how particular nonsymmetrical, indeed lumpy, components of the initial symmetrical quantum state of the universe become singled out in the course of the evolution of the universe. Where does this "collapse" come from if there is no external observer? This question has stimulated a range of possibilities, including a "no collapse" approach (Schlosshauer 2006) based on the subsequent development of Zeh's work by Zurek and others (Giulini et al. 1996) and the collapse-based theory of Penrose discussed in the next section. My aim in this chapter is to explain how these developments in quantum theory now give us a more integrated picture, differing both from Cartesian dualism and from scientific mechanism.

4.1.4   The Relation Between Physics and Consciousness

In the dualism of Descartes, there was a two-way relationship between mind and matter. Sense data were projected onto the pineal gland and thereby integrated, whence they were picked up by the soul. Reciprocally, the soul could move the

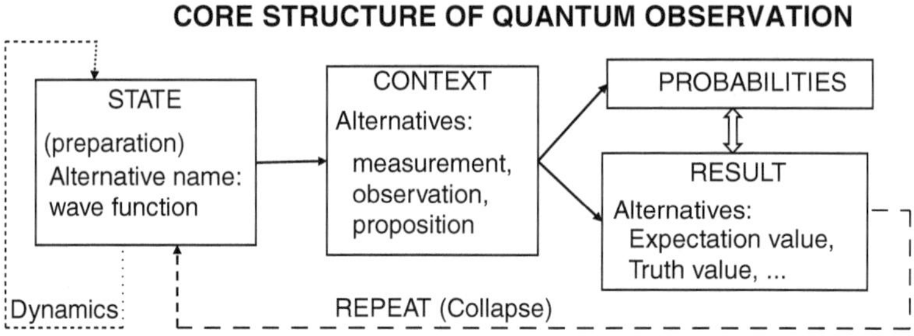

CORE STRUCTURE OF QUANTUM OBSERVATION

Fig. 4.1 Alternative terminologies for observations

pineal gland so as to steer the flow of spirit across the third ventricle of the brain and into appropriate nerves so as to execute the soul's purposes. Philosophically, the problem with this was not so much that the physiology was wrong (e.g., the pineal gland does not hang down into the third ventricle, and the nerves are not tubular channels) but that when it came to the details it was impossible to see how two metaphysical entities so different in kind could have anything to do with each other, even if one got the physiology right. *Res cogitans* was supposed to be independent of spatial position, so how could it be linked with the pineal gland of one particular organism? *Res extensa* was governed by strict physical laws (shortly to be expressed mathematically by Newton) so that the pineal gland required a force to move it; but how could one find room for an extra-physical force within the scheme of the newly emerging mechanics?

This problem seems to have driven the development of dual aspect approaches, as opposed to substance dualism. The first proponent of this was Spinoza, who made mind and matter distinct aspects of God. This was attractive metaphysically, but since God had no structure at the level of substance, there was still nothing to link the particularities of matter to the very different particularities of spirit. Similarly, Chalmers (1995) proposed that consciousness was an aspect of information, but without appropriate structures in each that links them together this would seem to be just as much an empty verbal formula as Spinoza's proposal.

The argument in this chapter breaks away from this problem through its quasi-Kantian approach. The dual poles of mechanical and ontological (Sect. 4.1.2.3) correspond to distinct ways of knowing, rather than substances or aspects of substances, and they are related through the reflexive capacity of our knowing, each being able to know the other; specifically, through each one's filling the gaps in the other. Ontology and mechanism are both required to explain how the universe behaves, because in quantum theory the universe is incompletely described without this. Consciousness actually *does* something (Sect. 4.2.4) in filling this gap so as to steer the universe. This is the "frontier of the mind–brain relationship." To articulate this, we need an integrated view of physics and consciousness – a profound challenge, but one toward which substantial steps have already been taken.

4.2   Toward an Integrated View of Physics and Consciousness

The main contender for such an integrated view is that of Hameroff and Penrose (1996) described in the next chapter. Below I critique some aspects of this, and then proceed to a variation of their scheme taking into account the developments in the concept of consciousness reviewed earlier.

4.2.1   Penrose's Model and an Alternative

The theory of Hameroff and Penrose derives from work definitively explored in Penrose (2004), which proposes a criterion for the collapse of the quantum state that must hold in any future unified physics. Collapse occurs if the state contains a super-position of elementary states whose difference has enough energy to create a single graviton (the hypothetical particle of quantum gravity). I will briefly list here the reasons why I am not convinced that Penrose and Hameroff's development of this thesis solves the consciousness problem, before going on to describe an alternative.

- The mechanism they describe is claimed to produce a nonalgorithmic information processing system in the brain. If correct, this would still not supply any connection with our immediate awareness, which is a matter of participating in experience, not processing information. Penrose's mechanism may explain why life is so clever, but not the "hard problem" (Chalmers 1995).
- In any case, there does not seem to be any reason why the collapse described by Penrose should involve information processing at all, rather than random selection. I would agree with him that humans are capable of nonalgorithmic information processing, but how does collapse actually achieve a nonalgorithmic processing of information (bearing in mind that quantum computers, as well as classical, are algorithmic)?
- The argument that gravitation produces collapse (Penrose 2004) needs a lot more investigation than it has so far received. Penrose makes an analogy between, on the one hand, the mathematics of gravitating superpositions which fail to have a stationary state and, on the other hand, the decay of an unstable nucleus: but the latter *presupposes* the occurrence of a collapse, so we cannot use this as an independent argument for collapse. The core of Penrose's argument is that when space-time loses its symmetry, it is not possible (in the semi-classical model implicitly being used) to superpose the states involved; but the failure of this model would seem to argue for the introduction of a fuller nonlinear theory, not for the appearance of a completely new collapse process intervening in quantum physics.

Here I will report on the development of an alternative based on the no-collapse approach mentioned earlier (Schlosshauer 2006). Many of the arguments which follow are equally applicable to the Penrose-Hameroff model: the ideas developed in the two approaches are not all mutually exclusive.

I will start with the ideas of Don Page (2001), who postulates that the quantum state of the universe as represented in standard cosmological models undergoes a dynamical evolution with no collapse. At any moment of time (time is well defined in the standard model, though it need not be in other models), the state can be expanded in any one of a vast numbers of ways as a superposition of other states. If one of the components of such a superposition has the properties which correspond to its being the state of a conscious entity, then the experience of this entity is a possible experience of a universe. So the possibilities for what the universe can look like to a conscious being are constrained both by the physics of the universe and by the properties that define consciousness. Consciousness imposes a selection criterion, or a perceptual filter, in the state of the universe. This is a quasi-Kantian move, as noted in Sect. 4.1.1: the world as we see it (as distinct from the hypothetical total universe of quantum physics) is one possible conscious snapshot that is selected out of the quantum state of the universe by criteria for consciousness. The double determination of the world, from both quantum physics and a selection effect imposed by the phenomenology of consciousness, is the essence of the version of no-collapse physics that I will be exploring below.

Page does not attempt to specify what it is about, say, a human brain that renders a system conscious, a specification that is essential for a practical application of the idea of consciousness as a filter. A first attempt at this was made by Donald (1990): unconvincingly, since it was based on a switching model for the brain that would apply equally well to a telephone exchange. In current science, the work of Hameroff, well grounded in physiological data, can be used as a basis for a selection model of quantum theory.

Before proceeding, I will summarize the interpretative positions on quantum theory that I have mentioned so far.

1. Dualistic (London and Bauer 1939, 1983): mind produces collapse.
2. No-collapse (Schlosshauer 2006): decoherence produces a classical (but statistical) world.
3. Enlargement of physics – Penrose (2004) and similar theories: a richer physical theory produces collapse.
4. Filtering by consciousness (Donald 1990; Page 2001): a criterion for the physical correlate of a conscious sensation selects possible observable universes from the quantum state without collapse.

Page (2001) uses the term "sensation" rather than "consciousness" (calling his proposal "mindless sensationalism"!) but his "sensation" is very close to the definition that I am using here of "consciousness." He applies the selection argument to just one moment of sensation by one conscious entity. If the conscious entity concerned has, in addition to consciousness, the sort of content that we call "memories," then this moment of awareness will seem to it as if it is part of a historical sequence of events. This is a rather extreme form of solipsism, where not only might I be the only conscious being in the universe, but my life might consist only of this one instant! I consider, however, that if consciousness is presencing, as noted above in Sect. 4.1.2, then it requires both other beings to be copresent with me and also requires myself to have been present in the past.

I will therefore modify Page's rather austere proposal to take this into account, by regarding the perceived world as the world of a community of conscious entities which are enduring in the sense of having repeated moments of awareness (or "sensations").

4.2.2   Histories and Cosmology

The extension of Page's approach to a succession of sensations involves a formalism known as the "histories interpretation" of quantum theory introduced earlier by Griffiths (1984), while the further extension to a concurrently existing collection of beings involves a formalism introduced by Hartle (1991), which he called "generalized quantum theory," following his joint work with Murray Gell-Mann in the early days of quantum cosmology. As neither of these is well-known outside physics circles, I shall describe these two steps further, though this will entail more detailed technical considerations in this section than elsewhere. I will begin with the histories interpretation.

A history, on this approach, is a succession of "observations," treated purely mathematically without inquiring who or what is doing the observation. Any single observation (subject to a few constraints relating to finiteness) is equivalent, in terms of the information delivered, to a set of mutually compatible binary observations (i.e., with a yes-or-no result) or propositions, which can be performed in any order. For example, the measurement of the position of a particle in a box 10-mm long to an accuracy of 1 mm can be reduced to ten propositions stating that the particle is in the first mm, the second mm, and so on. It is thus sufficient to restrict histories to sequences of (satisfied) propositions. In traditional quantum theory, this would be associated with a sequence of collapses of the wave function; the histories approach, however, notes that all that is required is a formula for evaluating the probabilities for the satisfaction of all such sequences of propositions, each occurring at a specified time, given some initial quantum state of the universe. Such a formula can be expressed without any reference to collapse.

While those who developed the histories formalism were not, as far as I know, concerned with the nature of sensation, the formalism lends itself well to this. A moment of awareness (an experience, in Page's terminology) can be thought of as a short time period in which an organism responds to the environment through a particular pattern of sensitivities. From a physics point of view, these are receptors and neural networks that are primed to respond to particular stimuli (Varela et al. 1993); from a subjective point of view, there are certain expectations, desires, and directions of attention. These sensitivities can be represented as propositions or questions, not necessarily conscious ones, which may or may not be affirmed (is the indistinct figure appearing in the distance someone I know? if so, who? Shall I move to greet them?), whose confirmation is registered in consciousness. Thus moments of awareness can, on a rather crude model, be associated with confirmed propositions, and a succession of such moments with a history of confirmed propositions.

The dependability and causal structure of the world as we see it is encoded in these probabilities. If, for example, my awareness a few seconds ago had included my sitting in my house, but in my awareness now the house has disappeared, the formalism would assign from this a very low probability factor to the overall tally.

The exercise of sensitivities leading to an awareness is associated both with a certain time duration (nothing is instantaneous, either in the brain or in consciousness) and with a certain location in space, namely the physical system in the organism that provides the locus for consciousness. In the language of relativity theory, it is associated with a region of space-time. Thus Hartle, working in the context of general relativity and cosmology, found it natural to extend the notion of a time-sequence of confirmed propositions to that of a collection of space-time regions associated with propositions based on many organisms, which we might call a "generalized history." Figure 4.2 gives a sketchy 2-dimensional flat idea of this 4-dimensional curved concept. The diamond shapes indicate sensations, the shape representing so-called "globally hyperbolic" regions of space-time, which have a maximum extent in time subject to their being causally dependent on a single instant within this duration (Clarke 2007; Hawking and Ellis 1973). The lines labeled "Space" and "Time" are intended to indicate the orientation of the diagram, but in accordance with relativity theory there are no distinguished directions of space and time. The diagonal lines emanating from sensation A indicate the regions lying wholly to the future or to the past of A. Sensations A and C are mutually dependent, contributing a factor to the formula for the probability (taking the place of collapse). This is not the case with A and C. Hartle called the approach stemming from this "generalized quantum theory."

Both the histories approach and generalized quantum theory require the specification of some initial quantum state. In Hartle's case, there is an obvious candidate for this, namely the homogeneous state of the universe in the earliest stages of its evolution. In this cosmological context, generalized quantum theory is vital in making sense of what is going on. The conventional interpretation of quantum theory involving collapse has to explain how a perfectly homogeneous state of the universe, subject to perfectly homogeneous physical laws, should produce a strongly inhomogeneous universe. The conventional answer is to appeal to "quantum fluctuations" in the early stage, but fluctuations arise from an observation of the state, and there are no observers present until they emerge from the very inhomogeneities that they are required to produce! John Wheeler (Clayton 2004; Wheeler 1994) bit the bullet of this paradox and proposed a reversed action in time. Generalized quantum mechanics reformulates the issue through a formula for a probability which involves both the emergence of sentience as a component of the developed quantum state and the registering of particular sensations by sentient beings.

As a side remark in support of Wheeler's idea, we can note that Dowker and Kent (1996) in an influential paper on the histories interpretation pointed out that instead of relating probabilities to past influences one could equally use influences from the future, or both future and past, a possibility that applies equally to generalized quantum theory. This may have applications to the study of abnormal experiences (Clarke 2008).

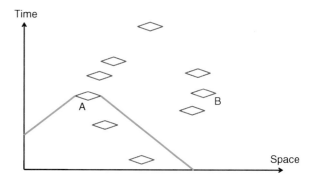

Fig. 4.2 Moments of awareness in space-time as a generalized quantum theory

When considered in more detail (Clarke 2007), the mathematical process for finding the probability p of a collection of moments of awareness gives practically the same final result as the "consciousness collapses the universe" position which I criticized in Sect. 4.1.3 – as is to be expected since they are simply reformulations of the same mathematical structures. The only numerical difference arises when moments of consciousness in different organisms occur more rapidly than the time taken by light to travel from one to another (i.e., outside the diagonal lines in Fig. 4.2). In that case conventional theory gives an answer involving communication that is "instantaneous" (a meaningless concept in a general cosmology) whereas Hartle's version gives a result that does not imply this. The important difference between the collapse interpretation and Hartle's is conceptual rather than mathematical: in a theory with collapse (positions 1 and 3 in the list in the previous subsection) we have to understand how consciousness produces a physical collapse process; in the approach of Page augmented by Hartle (position 4 in the list) we have to understand what is the physical correlate of consciousness (i.e., "what systems are conscious?") so as to apply this as a selection criterion on the states present in the wave function of the universe.

The role of the propositions that are used to model moments of awareness is central, so I will sketch briefly some further aspects. One of the main issues in quantum theory, an issue that was resolved by the discovery of decoherence (Sect. 4.1.3), is to understand why it is that we are never aware of states that are quantum superpositions of states that are grossly different at the macroscopic level: the famous "Schrödinger's cat" question of why we do not observe the state

$$state\_of\_cat = alive + dead$$

in circumstances where quantum theory predicts just this state. It turns out (Clarke 2001) that generalized quantum theory coupled with decoherence theory produces near zero probabilities for the occurrence of such an awareness in a generalized history. In other words, we need to use both positions 2 and 4 of the list above.

A further problem now arises, however. In classical logic, if I postulate a proposition P which experiment proves to be false, this constitutes a confirmation of the proposition not-P. This is the principle of the excluded middle: that there is in logic no "not proven" in between P and not-P. But in quantum theory the negation of the state *alive + dead* is the state *alive − dead*, which is just as impossible to include in a generalized history. Thus these values of probability violate the most basic rules of probability theory and/or classical logic, although they are more compatible with the nonstandard logics considered in Sect. 4.1.2.3.

The usual response to this is to rule out, a priori, histories that give rise to such a situation as "inconsistent" (the propositions involved do not in fact satisfy the logical rules of propositional calculus). More generally, the conventional histories interpretation restricts itself to "consistent histories," meaning ones where the values that I referred to as "probabilities" really behave like probabilities in the sense that the values for a set of propositions, which logically ought to be exclusive and covering all the possible cases, add up to 1. For large scale systems like Schrödinger's cat, we can interpret the zero probability of the states just mentioned as meaning that they are in fact not perceivable at all, as a result of the decoherence effect referred to in Sect. 4.1.3. We then experience a classical universe, as described in the monograph by Giulini et al. (1996), where these systems obey a classical logic. This need not, however, be the case on the borderline between large and small systems, where decoherence becomes weak. Then the possibility arises that the logic of sensations will obey a nonclassical logic (Clarke 2008) and the system involved (which we are supposing is a part of the brain of a sentient being) can behave as a quantum computer. This then gives a similar behavior to that in the approach to consciousness of Hameroff and Penrose (1996) (position 3).

4.2.3 What Things are Conscious?

The question arising from the Page-Hartle approach is, what are the criteria for consciousness? I have already severely limited the restrictions that these criteria can impose by defining consciousness as a *sui generis* quality distinct from the objective content of an entity's consciousness (which might include information processing) and also by claiming that we can only recognize consciousness through the nonverbal means of the relational subsystem. Thus it is hard, perhaps positively misguided, to draw a line between those entities that are conscious and those that are not. Several writers have recently considered this dilemma (de Quincey 2002; Mathews 2003; Skrbina 2005), and come to a panpsychist position: that everything is conscious. The meaning of "panpsychism" is ambiguous, however, varying between the idea that there is a diffused consciousness throughout the universe (a concept distinct from my usage here which refers to individual consciousness) and the idea, relevant here, that every "thing" is conscious.

This brings us to the question "what is a thing?" the title of a lecture series by Heidegger (1967) and, echoing this, the paper by Döring and Isham (2011) on ways of bypassing quantum collapse using topos logic. The former is to do with society's propensity for conceiving everything from the point of its utility, which is not the issue here. The latter is concerned with the process whereby the openness of the quantum state becomes channeled into a particular manifestation, but it is not concerned directly with consciousness, and would take us too far away from the current theme. In the context of recent supporters of panpsychism just cited, asking "what is a thing?" becomes "what structures are there within the basis of modern quantum physics that correspond to (comparatively) discrete individual entities, which include living organisms and organelles (noting that Hameroff locates consciousness at the microscopic level of organelles in the cell) and which can be discerned by the relational system?" I will shortly suggest a likely candidate for this, but I would stress that this is very tentative. The argument is at this stage leaving the area that has been well charted by experiment into a more speculative region, and my aim is to demonstrate that it is possible to single out at least one candidate for the loci of consciousness – to show that it is possible to bring together fruitfully the language of physics and the language of psychology.

One way into this is to ask instead, "what is not a thing?" For example "the set of everything green and displayed to view in my kitchen" – comprising, at a casual glance, a towel, a music poster, two cookery books, and a wiping cloth – is clearly an arbitrary collection of different "things" and not a thing itself. It is, however, not the fact of spatial separation in itself that disqualifies this set from thingness. Fred Hoyle wrote a plausible science fiction novel (Hoyle 1957) where the "invader from outer space" was a cloud of separate particles linked together only by magnetic fields. The introduction of fields (principally by Faraday) in classical physics opened a picture of the universe any of whose parts were interlinked by continuous physical structures, where separation was a matter of degree. And at the level of consciousness, I find it entirely meaningful from the relational standpoint to say that I am in some sense "one with" my wife, the Earth, etc. None of this holds with the collection of green objects just considered. Whitehead (1920) called such amalgamations "aggregates" and distinguished them from more organic collections that hung together as a whole.

In addition to "hanging together," it seems reasonable to require a thing to be a whole in the sense not being an arbitrary part of something greater. If, for example, I describe a brain at a certain level of detail as functioning as a whole, then a conceptual 1-cm thick slice of that brain will not be a whole. A thing needs to be maximal, as seen at a given level of detail. This issue of level of detail arises in considering nested things, such "me" as part of "me and my wife," a point that can be covered by decomposing states into tensor products of fine-grained and coarse-grained states (Clarke 2007).

At the present stage of this inquiry, it seems impracticable and unnecessary to attempt to define thingness in the full generality of panpsychism, but to try out some possible definitions that are at least broad enough to encompass the biological structures and quasi-biological structures (such as ecosystems) that we might consider as candidates for consciousness, in the nonfunctional sense of the term being used here. One tractable definition lies in *coherence* as the word is used by Mae-Wan Ho (Ho 1998), referring to a region (not necessarily connected spatially) that holds together dynamically, whose parts respond in harmony with each other.

Confusion can arise here from two distinct senses of "coherence" in the literature. One is the opposite of "decoherence" (Sect. 4.1.3): a decohered region has separate states in separated parts, so that the total state is just the tensor product of the states in its parts. Coherence in the sense of the opposite to this occurs when the states of all separated parts are entangled. In this state, events in different parts are correlated without their being any communication between them, as required in Mae-Wan Ho's concept. The other sense of "coherent" occurs in optics, including quantum optics, where the phases of the light in two separate regions are in step. But this can happen coincidentally without any entanglement between the two parts – as when, for instance, beams of light from two different lasers set up at exactly the same frequency are brought together and produce an interference pattern (Magyar and Mandel 1963). This sense does not capture the idea of holding together dynamically. My proposal for a "thing," therefore, is a local quantum state all of whose spatially separated parts are entangled (I shall call such a state coherently entangled) and which is maximal in the sense of not being a spatial part of a larger state that is also coherently entangled. While the concept is quite intuitive, the actual definition of the degree of coherent entanglement, taking into account the existence of nested things, is rather complex (Clarke 2007: A2.3). Importantly, this definition covers exactly the situation required for microtubules in the approach of Hameroff and Penrose; so we are considering the same sort of entity, but from somewhat different viewpoints. Importantly, we are considering systems that are essentially quantum mechanical, even though the ideas may be extensible to more general "things."

4.2.4   What Does Consciousness Do?

Though it might turn out that consciousness, as defined here, is an epiphenomenon, a passive by-product of the physical aspect of our brain, our subjective experience is that consciousness does something: we not only enjoy our world (Alexander 1920), but we feel that through responding consciously we are agents in shaping it. That is, in some sense we have "free will." This idea is clear cut in Cartesian substance dualism: with a few caveats, either the soul acts on the body (as well as the body acting on the soul) or it does not. When we move away from this, however, free will and related ideas become very elusive (e.g., McFee 2000; Van Inwagen 1983). In the present context, where we are talking about consciousness as a subjective aspect of a physical system, the question is whether or not the overall behavior of the system is discernibly different in the presence of consciousness as compared with the absence of consciousness. Here, moreover, "discernibly" must for consistency include what can be discerned purely through our relational faculties, while for nontriviality "behavior" must go beyond the mere reflexive examination of changes in an individual consciousness by itself.

Another complication when thinking about free will and related issues stems from the results of Batthyany (2009) on extending Libet and Soon's work on the neurology of decision making. It becomes clear from this that in most everyday circumstances decision making lies not with the explicit consciousness (Sect. 4.1.2.2) but with the other loci of consciousness which, in relation to the explicit "I" we lump together as the "unconscious." It is, however, likely from introspective evidence that these different loci of consciousness are interlinked at the level of consciousness, even if not at the physical level. With these rather general considerations in mind, I will move on to what consciousness plausibly can and cannot "do."

Since we are considering quantum systems rather than classical systems (because of the choice in the previous section of the category of possibly conscious "things"), and since such systems are famously indeterminate, it might be thought that consciousness could act by forcing one outcome of a quantum measurement that otherwise would be entirely random. In terms of the generalized quantum theory picture used here (Sect. 4.2), this is equivalent to consciousness modifying the probabilities p that quantum theory associates to propositions P at any of the moments of awareness in space-time. There are, however, two objections to this. First, the only element in generalized quantum theory which depends on the basic dynamical laws of physics is the evaluation of these probabilities. If every conscious system in the universe could tinker with this dynamic, it would make nonsense of the whole concept of physical law. The second objection is a more specific aspect of the first: this simple approach is ruled out by experiments on the parapsychological phenomenon of "psychokinesis." If, at the basic level of everyday action and decision making, consciousness were to modify these probabilities in a way that was at least indirectly linked with the explicit "I," then one would expect that we could easily and regularly influence events that depend on quantum mechanical events such as radioactive decay. But this is just the possibility that is examined in many experiments in parapsychology. In fact it turns out that, in the terminology of statistics, this is at best an extremely weak, albeit clearly significant effect (Radin 1997, Chap. 8), ruling out the modification of probabilities by consciousness.

What consciousness can do without contradicting current experiment, however, is to modify not the probabilities p but rather which propositions P are associated with each moment of awareness. In the original quantum theory, the Ps were given by the laboratory context. When physicists extended their view to the cosmos and decoherence was discovered, "observation" was identified with massive "subsystems" of the cosmos in interaction with their environment which defined a unique set of Ps (which Zurek termed the "pointer basis") that could happen in any order. This assumed, however, that we were "given" a criterion for what a subsystem of the cosmos is, which is the same sort of question as what a "thing" is. Here we are considering coherently entangled states: regions that are on the borderline between completely quantum systems like atoms and completely classical systems dominated by decoherence. These are large enough for some propositions to lead to perceptions that are to some extent stable but not so large that there is a unique set of propositions singled out. This is the realm in which a system is open to consciousness acting as a noncausal influence, and hence in a way that does not contradict the well tested causal laws of physics.

The process whereby consciousness influences the propositions that are present at each moment involves subjective awareness rather than purely logical information processing. It is the "bare fact of awareness," which according to McGilchrist and Barnard (Sect. 4.1.2.2 above) is primarily linked to the relational way of knowing, and which can be understood through our "theory of mind" that is also rooted in our relational system rather than our rational system. Relational processes involve categories such as "meaning" (the relationships that different elements within consciousness have to one another) and "salience" (the extent to which the being of an element is present to other elements). Although only fully accessible to relational knowing, the rational way of knowing can approximate this by using nonstandard logics such as biologic (Matte Blanco 1998) and context-dependent logic (Goddard and Routley 1973; Routley et al. 1982). These logics link into the dynamics of quantum theory when the latter is reformulated in terms of topos logic, as has been achieved by Isham (Döring and Isham 2011) and others.

From the relational point of view, consciousness can bring to the world not only a subjective "what the world is like," but also a qualitative "what the world could be," which shapes the propositions that are present. When approximated rationally in terms of a proposition P, I shall call this *asserting* P. This relational process, on the one hand, and the rational dynamics of quantum theory expressed through the probabilities derived from generalized quantum theory, on the other hand, operate in a complementary fashion. Quantum theory has an openness, an incompleteness, in terms of what propositions are asserted at each moment of awareness. The play of meaning and salience within consciousness has an openness to what is given as physical law. The world comes into being as a result of this dual process.

4.3   Summary, Conclusions, and Future Prospects

To summarize, the developments that I have reviewed across several subject areas have built up to a radical and comprehensive break with the Cartesian world view.

In philosophy the "Copernican revolution" of Kant, although at first tied to categories that supported a Newtonian viewpoint, led to a steadily deepening realization (progressing a winding course through Hegel, Husserl, Heidegger, and Jorge Ferrer) that the world could only be understood by combining noetic and physical aspects. (As an Englishman, however, I must acknowledge an important alternative strand opposed to "continental philosophy.")

In psychology and neurophysiology, the idea of mind as a single discrete conscious cognitive entity became increasingly untenable. Though the particular structure and origin of Freud's "unconscious" remains contested, his introduction of cognitive processes alongside the conscious mind, at first regarded as absurd, has now become self evident, while our conception of the brain as a collection of rigid machines has changed to a picture of a plastic multiply connected network of shifting interactions, with a corresponding shift in our conception of the mind.

In physics, the Newtonian world at first changed to a deterministic machine whose parts interact through local contact. The advent of quantum mechanics, however, transformed this once again into a system that was nonlocal, nondeterministic, and inherently open to additional nonmechanistic processes. At the same time, the development of systems theory showed that nonpredictability and the emergence of novelty were equally a part of classical mechanics. A universe that is "mechanical" in the classical sense of the word has become untenable.

In logic the millennia-long project of founding logic on unique self-evident principles foundered on Russell's paradox, Gödel's incompleteness theorems and the flowering of numerous alternative logical schemes through the work of Prior, Robinson, Brouwer, and many others, open up alternative logics.

I should add that, a fortiori, these developments also close off materialism, which is merely Cartesianism with the soul removed, and so the modern resurgence of materialism is surprising. As Saulo Araújo remarks in his chapter in this volume (see Chap. 1), "It is as if the entire period of our intellectual history that goes from the second half of the eighteenth century until the late nineteenth century had not existed, so that this new materialistic dawn may sound like something really new."

My aim in the preceding section has been to bring together work that indicates one natural way forward into a post-Cartesian paradigm. The starting point is the argument of Nagel and Chalmers that the simple erasure of mind from the Cartesian picture untenably leaves us no experienced world whatever; it ignores, moreover, the whole thrust of the historical developments just sketched. The work of Teasdale and Barnard and McGilchrist, however, indicates the effectiveness of the next most simple system: a unitary model supporting a duality of function. I hope to have shown that there already exists enough work to constitute a solid account of such a system, to which I have added, as working hypotheses, the idea of spatial coherence as a determinant of the loci of consciousness.

This then opens up a range of present and future research areas for post-Cartesian science, developed in detail elsewhere (Clarke 2007).

1. A locus of consciousness can maintain its own coherence by repeatedly asserting its own internal entanglement, employing the well established quantum Zeno effect (Leibfried et al. 2003). This implements Spinoza's "conatus" stressed by Mathews (2003) as an ingredient of pan-psychism. This in turn makes possible a larger locus of consciousness that would otherwise be possible, an important consideration in the theory of Hameroff and Penrose (1996).
2. In addition, where the loci of consciousness of two people become momentarily entangled, the same Zeno mechanism can maintain the entanglement, establishing a correlation between the actions of the people (though not necessarily involving the transfer of classical information).
3. Where quantum computation is taking place, consciousness can assert projections on superpositions of classically incompatible large-scale situations, enabling creative insights impossible in conventional logical processes.
4. As noted above (Sect. 4.2.2), Hartle's process is symmetric in time, implying that future assertions can affect the present. This produces apparently "parapsychological" effects. This can radically alter the way we think about data such as apparent reincarnation or the factual content of near death experience (though not necessarily their spiritual content).

A great many aspects of this approach remain open, awaiting experimental investigation. The approach has much in common with that of Hameroff and Penrose, and so experimental work of the kind described elsewhere in this issue by Hameroff can both refine the present theory and also distinguish between the two theories. The following experimental questions seem critical in this respect.

5. Are some biological systems specifically structured so as to enhance internal coherence as defined above?

6. If so, is the decoherence time influenced by gravity (Penrose) or environmental interaction (Giulini et al. 1996)?

In addition, this line of investigation provides, for perhaps the first time, a possible theoretical framework (Clarke 2007) for so-called parapsychological effects, an area still often regarded as unscientific because the lack of such a framework, but one which could in principle discriminate between alternative theories of consciousness.

References

Alexander, S. (1920). *Space, time, deity*. London: Macmillan.

Batthyany, A. (2009). Mental causation and free will after Libet and Soon: Reclaiming conscious agency. In A. Batthyany & A. C. Elitzur (Eds.), *Irreducibly conscious: Selected papers on consciousness* (p. 135ff). Heidelberg: Winter.

Chalmers, D. J. (1995). Facing up to the problem of consciousness. *Journal of Consciousness Studies, 2*, 200–219.

Clarke, C. J. S. (2001). The histories interpretation: Stability instead of consistency? *Foundations of Physics Letters, 14*(2), 179–186.

Clarke, C. J. S. (2007). The role of quantum physics in the theory of subjective consciousness. *Mind and Matter, 5*(1), 45–81.

Clarke, C. J. S. (2008). A new quantum theoretical framework for parapsychology. *European Journal of Parapsychology, 23*(1), 3–30.

Clarke, C. J. S., & King, M. (2006). Laszlo and McTaggart – in the light of this thing called Physics, *Network Review* (pp. 6–11). Winter 2006.

Clayton, P. D. (2004). Emergence: Us from it. In J. D. Barrow, C. W. Davies, & C. L. Harper Jr. (Eds.), *Science and ultimate reality* (pp. 577–606). Cambridge: Cambridge University Press.

de Quincey, C. (2002). *Radical nature: Rediscovering the soul of matter*. Montpelier: Invisible Cities Press.

Donald, M. (1990). *Quantum Theory and the Brain, Proceedings of the Royal Society Series A, 427* (pp. 43–93). London

Döring, A., & Isham, C. (2011). "What is a thing?" Topos theory in the foundations of physics. In B. Coecke (Ed.), *New structures for physics, lecture notes in physics* (Vol. 813, pp. 753–941). Berlin: Springer.

Dowker, F., & Kent, A. (1996). On the consistent histories approach to quantum mechanics. *Journal of Statistical Physics, 82*, 1575–1646.

Ferrer, J. N. (2002). *Revisioning transpersonal theory*. Albany: State University of New York Press.

Giulini, D., Joos, E., Kiefer, C., Kupsch, J., Stamatescu, I.-O., & Zeh, H. D. (1996). *Decoherence and the appearance of a classical world*. Berlin: Springer.

Goddard, L., & Routley, R. (1973). *The logic of significance and context* (Vol. 1). Edinburgh: Scottish Academic Press.

Griffiths, R. B. (1984). Consistent histories and the interpretation of quantum mechanics. *Journal of Statistical Physics, 36*, 219–272.

Hameroff, S., & Penrose, R. (1996). Conscious events as orchestrated space-time selections. *Journal of Consciousness Studies, 3*(1), 36–53.

Hartle, J. (1991). The quantum mechanics of cosmology. In S. Coleman, P. Hartle, T. Piran, & S. Weinberg (Eds.), *Quantum cosmology and baby universes*. Singapore: World Scientific.

Hawking, S. W., & Ellis, G. F. R. (1973). *The large scale structure of space-time*. Cambridge: Cambridge University Press.

Heidegger, M. (1967). *What is a Thing?* (W. B. Barton Jr. & V. Deutsch, Trans.). Chicago: Henry Regnery Company.

Ho, M.-W. (1998). *The rainbow and the worm: The physics of organisms* (2nd ed.). Singapore: World Scientific.

Hoyle, F. (1957). *The black cloud*. London: William Heinemann.

Leibfried, D., Blatt, R., Monroe, C., & Wineland, D. (2003). Quantum dynamics of single trapped ions. *Reviews of Modern Physics, 75*, 281–324.

Lockwood, M. (1989). *Mind brain and the quantum: the compound 'I'*. Oxford: Blackwell.

London, F., & Bauer, E. (1939). *La théorie de l'observation en mécanique quantique*. Paris: Hermann.

London, F., & Bauer, E. (1983). The theory of observation in quantum mechanics (translation of the above). In J. A. Wheeler & W. H. Zurek (Eds.), *Quantum theory and measurement* (pp. 217–259). Princeton: Princeton University Press.

Magyar, G., & Mandel, L. (1963). Interference fringes produced by superposition of two independent maser light beams. *Nature, 198*, 255.

Mathews, F. (2003). *For love of matter: A contemporary panpsychism*. Albany: State University of New York Press.

Matte Blanco, I. (1998). *The unconscious as infinite sets: An essay in bi-logic*. London: Karnac Books.

McFee, G. (2000). *Free will*. Teddington: Acumen.

McGilchrist, I. (2009). *The master and his emissary: The divided brain and the making of the Western world*. New Haven: Yale University Press.

Nagel, T. (1974). What is it like to be a bat? *Philosophical Review, 83*(4), 435–450.

Page, D. N. (2001). Mindless sensationalism: A quantum framework for consciousness. In Q. Smith & A. Jokic (Eds.), *Consciousness: New philosophical essays* (pp. 468–506). Oxford: Oxford University Press.

Penrose, R. (2004). *The road to reality: A complete guide to the laws of the universe* (pp. 846–856). London: Jonathan Cape.

Premack, D. G., & Woodruff, G. (1978). Does the chimpanzee have a theory of mind? *The Behavioral and Brain Sciences, 1*, 515–526.

Radin, D. (1997). *The conscious universe*. New York: HarperCollins.

Routley, R., Plumwood, V., Meyer, R. K., & Brady, R. T. (1982). *Relevant logics and their rivals, part 1, the basic philosophical and semantical library*. Atascadero: Ridgeview.

Savile, A. (2005). *Kant's critique of pure reason: An orientation to the central theme*. Malden: Blackwell.

Schlosshauer, M. (2006). Experimental motivation and empirical consistency in minimal no-collapse quantum mechanics. *Annals of Physics, 321*, 112–149.

Skrbina, D. (2005). *Panpsychism in the west*. Cambridge: Bradford Books.

Squires, E. J. (1994). Quantum theory and the need for consciousness. *Journal of Consciousness Studies, 1*(2), 201–204.

Steiner, G. (1989). *Real presences*. London: Faber and Faber.

Teasdale, J. D., & Barnard, P. J. (1993). *Affect, cognition and change: Remodelling depressive thought*. Hove: Lawrence Erlbaum Associates.

van Atten, M. (2009). The development of intuitionistic logic. In N. Edward Zalta (Ed.), The Stanford Encyclopedia of Philosophy (Summer 2011 Edition). Retrieved, from http://plato. stanford.edu/archives/sum2009/entries/intuitionistic-logic-development/.

Van Inwagen, P. (1983). *An essay on free will*. Oxford: Clarendon.

Varela, F. J., Thompson, E., & Rosch, E. (1993). *The embodied mind*. Cambridge: MIT Press.

Velmans, M. (2000). *Understanding consciousness*. London: Routledge.

Wheeler, J. A. (1994). *At home in the universe*. New York: AIP.

Whitehead, A. N. (1920). *The concept of nature*. Cambridge: Cambridge University Press.

Zeh, H. D. (1970). On the interpretation of measurements in quantum theory. *Foundations of Physics, 1*, 69–76.

Chapter 5
The "Quantum Soul": A Scientific Hypothesis

Stuart Hameroff and Deepak Chopra

Abstract The concept of consciousness existing outside the body (e.g. near-death and out-of body experiences, NDE/OBEs, or after death, indicative of a 'soul') is a staple of religious traditions, but shunned by conventional science because of an apparent lack of rational explanation. However conventional science based entirely on classical physics cannot account for normal in-the-brain consciousness. The Penrose-Hameroff 'Orch OR' model is a quantum approach to consciousness, connecting brain processes (microtubule quantum computations inside neurons) to fluctuations in fundamental spacetime geometry, the fine scale structure of the universe. Recent evidence for significant quantum coherence in warm biological systems, scale-free dynamics and end-of-life brain activity support the notion of a quantum basis for consciousness which could conceivably exist independent of biology in various scalar planes in spacetime geometry. Sir Roger Penrose does not necessarily endorse such proposals which relate to his ideas in physics. Based on Orch OR, we offer a scientific hypothesis for a 'quantum soul'.

5.1 Brain, Mind, and Near-Death Experiences

The idea that conscious awareness can exist after bodily death, generally referred to as the "soul," has been inherent in Eastern and Western religions for thousands of years. In some traditions, memories and awareness may be transferred after death to other lifetimes: reincarnation. In addition to beliefs based on religion, innumerable

S. Hameroff (✉)
Departments of Anesthesiology and Psychology, Center for Consciousness Studies,
The University of Arizona Medical Center, 1501N Campbell Ave, Tucson, AZ 85724, USA
e-mail: hameroff@u.arizona.edu

D. Chopra
The Chopra Center, 2013 Costa del Mar, Carlsbad, CA 92009, USA
e-mail: rishi@chopra.com

A. Moreira-Almeida and F.S. Santos (eds.), *Exploring Frontiers of the Mind-Brain Relationship*, Mindfulness in Behavioral Health, DOI 10.1007/978-1-4614-0647-1_5, © Springer Science+Business Media, LLC 2012

subjects have reported conscious awareness seemingly separating from the subject's brain and physical body; this occurs in conjunction with so-called near death experiences (NDEs), most typically in patients who have been resuscitated after cardiac arrest (e.g., van Lommel et al. 2001; Parnia et al. 2007). Such patients describe remarkably consistent phenomenology including visions of a white light, being in a tunnel, feelings of serenity, conversing with deceased loved ones, life review and, in some cases, floating out of the body (out-of-body experiences – OBEs). Frequently, NDE/OBE patients also report a subsequent loss of the fear of death, and tend to be more serene and accepting of life's vicissitudes (Chopra 2006).

Somewhat comparable experiences have been reported in various types of meditative and altered states, as well as traumatic psychological events, or seemingly without cause. A Gallup poll estimated some ten million Americans have reported some form of NDE/OBE (Chopra 2006). The drug ketamine, used as a "dissociative" anesthetic, can produce subjective reports of conscious awareness outside the body (Jansen 2000), as can various other psychoactive drugs. But subjective reports of drug-induced effects are distinctly different from those of NDEs/OBEs (Greyson 1993).

Unable to explain NDEs/OBEs, modern science on the whole ignores and derides such reports as unscientific folly, illusions due to stimulation of particular brain regions (Blanke et al. 2004), or hallucination due to hypoxia (lack of oxygen; Blackmore 1998). But in response one can point out: (1) subjective reports of illusions of body image are quite limited and completely different from NDE/OBE descriptions, (2) hypoxic patients are agitated, not serene, and do not form memory, and (3) modern science cannot explain normal, in-the-brain consciousness.

This last point is critical. NDEs/OBEs are particular types of subjective conscious awareness, in some way akin to our everyday conscious experience (including dreams). How the brain produces consciousness remains unknown.

The prevalent modern scientific approach to consciousness casts the brain as a biological computer, with 100 billion neurons and their axonal firings and synaptic connections acting as information networks of "bit" states and switches. Variability in synaptic strengths mediated by chemical neurotransmitters shapes network activity and enables learning and intelligent functions (Hebb 1949; Crick and Koch 2001; 2004). This "brain-as-computer" view is able to account for complex nonconscious cognitive functions including perception and control of behavior. Such nonconscious cognitive functions are described as "zombie modes," "auto-pilot," or "easy problems" (Koch and Crick 2001; Hodgson 2007; Chalmers 1996). The "easiness" derives from the apparent cause-and-effect between specific computational functions of brain neurons, and actions and behavior which do not involve conscious will or phenomenal experience.

The "hard problem" (Chalmers 1996) is the question of how cognitive processes are accompanied or driven by phenomenal conscious experience. Despite detailed understanding of neuronal firings, synaptic transmissions, neurotransmitter chemistry, and neuronal computation, there is no accounting for conscious experience, the "self," free will or "qualia" – the essence of experienced perceptions. How can the redness, texture, and fragrance of a rose, the experiential world, derive from data streams and electrochemical activity?

The answer according to most views in modern science is that consciousness emerges from a critical (but unspecified) level of neuronal computational complexity. In nonlinear dynamics, new properties do emerge in hierarchical systems, but such systems abound in nature and technology without consciousness. (e.g., weather patterns, the internet). The notion that computational complexity *per se* can account for consciousness may be mere wishful thinking.

The brain-as-neuronal-computer view has three problems.

1. Because brain synaptic computation correlating with sensory processing often occurs after we have responded to that sensory input (seemingly consciously), the conventional view in modern science is that consciousness occurs after-the-fact, and that conscious control is an illusion, consciousness is merely along for the ride (Dennett 1991; Wegner 2002). Apparently we are, as T.H. Huxley (1893) famously said, "helpless spectators".
2. The best measurable correlate of consciousness (gamma synchrony EEG) does not derive from synaptic computation. Synchronized electroencephalography (EEG) in the gamma range of 30–90 cycles per second (Hertz, "Hz") occurs in various brain regions at different times concomitant with consciousness (Gray and Singer 1989a,b; Engel et al. 1991; Singer 1995; 1999). Gamma synchrony requires networks of neurons interconnected not only by axon-to-dendrite chemical synapses, the basis for recognized neuronal computation, but by dendrite-to-dendrite gap junction electrical synapses (Christie and Westbrook 2006; Dermietzel 1998). One unconventional view is that gap junctions in various neurons open and close, enabling mobile zones of gamma synchrony to move about the brain, mediating consciousness (Hameroff 2006; 2010).
3. As cells, neurons are far more complex than simple switches. Consider the unicellular *Paramecium* which can swim around, find food and mates, avoid obstacles, learn and have sex, all without a single synaptic connection. Artificial intelligence (AI) efforts to simulate brain function have yet to simulate anything as intelligent and nimble. *Paramecium* utilizes intelligent organizational functions of cytoskeletal lattice polymers called microtubules (Sherrington 1953). These same microtubules form the internal structure of brain neurons, regulate synapses and disintegrate in Alzheimer's disease (e.g. Brunden et al. 2011). Microtubule information processing may underlie neuronal function.

Unable to explain consciousness in the brain, conventional science ignores apparent evidence for NDEs/OBEs, rejecting even the possibility of their occurrence. There are, however, unconventional but scientifically valid approaches to consciousness, which may address the three problems described above, and accommodate NDEs/OBEs as well as possible conscious awareness after bodily death. Such approaches explore strata of nature at an even finer scale than the chemical reactions and electrical signals relied upon by neuroscience, seeking convincing answers at the quantum level instead.

5.2   The Quantum World and Fine Scale of the Universe

Quantum theory tells us that physical processes occur in discrete, quantized steps, or levels. The laws that govern the quantum differ strangely from the predictable reality of our everyday "classical" world. At small scales, and sometimes at large scales, the bizarre laws of quantum mechanics reign. For example, atoms and sub-atomic quantum particles can exist in two or more states or places simultaneously, more like waves than particles, and existing as multiple coexisting possibilities known as *quantum superposition,* governed by a quantum wave function. Another quantum property is "nonlocal entanglement," in which components of a spatially separated system remain unified and connected (Penrose 1989).

Physics circumvents the strangeness of quantum mechanics by strictly dividing the macro/classical and micro/quantum, keeping the two worlds apart. However, consciousness somehow bridges the macro/classical and micro/quantum domains, equivalent to the subject – object split. Consciousness exists precisely on the edge between quantum and classical.

In our conscious experience, we do not see superpositions – coexisting wave-like possibilities. We see objects and particles as material things in specific locations and states. This is partly due to scale. A humpback whale leaps out of the sea whole, despite the fact that the atoms and subatomic particles comprising the whale may occupy uncertain or even multiple positions in the invisible realm of possibilities. But even when small quantum systems are measured or observed they somehow choose definite states.

The issue of why we do not see quantum superpositions in our everyday classical world is known as the "measurement problem," which has led to various interpretations of quantum mechanics. Early experiments by quantum pioneer Niels Bohr and others seemed to show that quantum superpositions, when measured by a machine, stayed as multiple possibilities until a conscious human observed the results. Bohr concluded that conscious observation "collapsed the wave function," that unobserved superpositions persisted until being observed, at which instant they reduced, or collapsed to particular definite states (the choice of states being random). In this approach, consciousness *causes* quantum state reduction, placing consciousness outside science. Erwin Schrödinger objected through his still-famous thought experiment in which the fate of a cat in a box is linked to a quantum superposition. According to the Copenhagen interpretation (so-named after Bohr's Danish origin) Schrödinger's cat is both dead and alive until the box is opened and the cat observed. The thought experiment was intended to ridicule Copenhagen, but the question remains: how large can superpositions become?

Another popular interpretation is the multiple worlds view (Everett 1957) in which superpositions are separations in reality, each possibility evolving its own distinct universe; a multitude of coexisting universes results.

Yet another approach is decoherence in which interaction with classical world erodes quantum states. But decoherence does not address isolated quantum systems. Finally, various types of objective reduction (OR) propose that specific objective thresholds cause quantum state reduction.

One particular OR theory was proposed by British physicist Sir Roger Penrose (1989), who began by addressing the fundamental nature of superposition. He extended Einstein's general theory of relativity, in which matter is essentially space-time curvature, to the Planck scale (10^{-33} cm), the most basic level of the universe. A particle in one state or location would be a specific curvature in space-time geometry, and the same particle in another location would be curvature in the opposite direction, extending downward to the Planck scale. Superposition of both locations can then be seen as simultaneous curvatures in opposite directions, and hence, according to Penrose, a separation, bubble, or blister in the very fabric of reality.

If such space-time separations were to continue and evolve, the universe would bifurcate, leading to parallel universes as described in the multiple worlds view supported by many physicists and cosmologists including Stephen Hawking (Hawking and Mlodinow 2010). But Penrose has suggested that such space-time separations are unstable and will reduce, or collapse to one particular state or location at a particular time due to an objective threshold intrinsic to the fine structure of the universe, like infinitesimally tiny soap bubbles bursting one facet or another, shaping and creating a new reality. Penrose also suggests that each OR, or self-collapse – essentially a ripple or quantized annealing in fundamental space-time geometry – results in a moment of conscious experience.

This is in direct contradistinction to the Copenhagen interpretation in which consciousness is outside science, externally *causing* reduction by observation. In Penrose OR, consciousness *IS* reduction (a particular type of reduction). Thus Penrose OR is the only worldview incorporating consciousness into the universe.

Penrose OR differs in another important way from Copenhagen and decoherence in which particular classical states are selected randomly from among superpositioned possibilities. The selections in Penrose OR are not random, but influenced by information embedded in fundamental space-time geometry, information Penrose characterized as Platonic values (Penrose 1989).

The Greek philosopher Plato described an abstract world of pure form, mathematical truth, and ethical and aesthetic values. Penrose suggests such Platonic values, along with precursors of physical laws, constants, forces, and consciousness, literally exist as patterns in fundamental space-time, encoded in Planck scale geometry.

Physics tells us the universe is as it is, and thus able to support life and consciousness, because 20 or more physical constants and the laws they dictate take on very specific values. If any of these varied even slightly, we would not be here, so the precise values and our presence in the universe are apparently a coincidence of staggeringly low probability, akin to wining the cosmic lottery. The "anthropic principle" addresses the question of why these values are what they are, and has several interpretations (e.g., Davies 2006). The most common is tautological – that we are in the particular universe which has these specific values simply because it *has* those values. If it did not, we would not be here. For many physicists and philosophers, the tautological answer is related to the multiple worldviews, that this universe with consciousness is one in a multitude of universes, the others having different physical constants and lacking life and consciousness.

This is the view espoused by Hawking and Mlodinow in their book *Grand Design* (Hawking and Mlodinow 2010) in which they assert "M-theory" (a derivative of string theory) with a near-infinite number of parallel universes, all others devoid of consciousness.

Penrose suggests another possibility which avoids the need for multiple universes. Values for physical constants defining our universe may be encoded in the fine structure of the universe itself, along with mathematical truth, Platonic values, and precursors of mass, spin, charge, and consciousness. The roots of consciousness may thus extend to the most basic level of the universe. Penrose has also proposed that our universe is serial, that the Big Bang was preceded by a previous iteration, and before that another one and so on (Penrose 2010). Unlike the idea of parallel universes which is untested (and likely untestable), the Penrose proposal for serial universes is supported by evidence from the cosmic microwave background radiation (Gurzadyan and Penrose 2010). Perhaps physical constants, conscious precursors, and Platonic values embedded in the fine structure of the universe mutate and evolve with each cosmological cycle.

What *is* the fine structure of the universe? The material world is composed of atoms and subatomic particles. But atoms ($\sim 10^{-8}$ cm) are mostly empty space, as is the space between atoms. If we go down in scale from atoms, eventually we reach the basement level of reality, Planck scale geometry at 10^{-33} cm, with coarseness, irregularity, and information.

Descriptions of Planck scale geometry include string theory and loop quantum gravity. String theory, in which Planck scale strings vibrate at specific frequencies correlating with fundamental particles, has several problems. It lacks background geometry (e.g., in which the strings vibrate) and requires multiple untestable dimensions (Penrose 2004).

Another approach, loop quantum gravity depicts space-time geometry as quantized into volume pixels, Planck scale polygons whose edges may be considered as irreducible spin whose lengths also vary but average 10^{-33} cm. Planck volumes evolve and change with time, conveying information as a 3-dimensional spider web of spin. Somehow, space-time geometry is also nonlocal, as revealed by entanglement experiments (Nadeau and Kafatos 2001), and perhaps holographic (e.g., Susskind 1994). Could Planck scale information affect biology?

Recent evidence suggests that Planck scale information may repeat at increasing scales in space-time geometry, reaching to the scale of biological systems. The British-German GEO 600 gravity wave detector near Hanover, Germany has consistently recorded fractal-like noise which apparently emanates from Planck scale fluctuations, repeating every few orders of magnitude in size and frequency from Planck length and time (10^{-33} cm; 10^{-43} s) to biomolecular size and time (10^{-8} cm; 10^{-2} s, Hogan 2008; Chown 2009). At some point (or actually at some complex edge, or surface) in this hierarchy of scale, the microscopic quantum world transitions to the classical world. If this transition is due to Penrose OR, consciousness occurs as a process on this edge between quantum and classical worlds.

This notion that consciousness is in some way intrinsic to the universe is comparable to purely subjective views on consciousness going back thousands of years in India. The Vedic tradition and ancient sacred texts derive their name from the Sanskrit word *Veda,* for knowledge. The most philosophical branch of Veda is Vedanta – literally, "the end of the Vedas." In Vedanta, consciousness is everything, and manifests, or creates reality. In this view (taken by one of us, DC, differing slightly from the argument presented in this paper), consciousness is both subject and object, both quantum and classical. Consciousness is all there is (Chopra 2001).

Penrose OR (and Penrose-Hameroff Orch OR) maintains the classical world exists on its own. Consciousness is a process on the edge between quantum and classical worlds, the process consisting of discrete, quantized ripples in the fine scale structure of the universe, transitions between subject and object.

5.3 Quantum Consciousness: Orchestrated Objective Reduction ("Orch OR")

The Penrose-Hameroff theory of "orchestrated objective reduction" ("Orch OR") proposes that consciousness depends on quantum computations in structures called microtubules *inside* brain neurons, occurring concomitantly with and supporting neuronal-level synaptic computation (Penrose and Hameroff 1995; Hameroff and Penrose 1996a,b; Hameroff 1998a,b; Hameroff et al. 2002).

Microtubules are cylindrical polymers of the protein "tubulin," and major components of the cell cytoskeleton which self-assembles to configure intracellular architecture, create and regulate synapses, and communicate between membrane structures and genes in the cell nucleus. In addition to bone-like support, microtubules and other cytoskeletal components seem to act as the cell's nervous system, its "on-board computer," continually reshaping and differentiating. In microtubule lattices, states of individual tubulins are proposed to act as "bit" states, as in classical computers and molecular automata (Hameroff and Watt 1982; Rasmussen et al. 1990). Microtubule-level processing raises the capacity for neuronal information processing immensely. Rather than a few (synaptic) bits per neuron per second, 10^8 tubulins per neuron switching coherently in megahertz (10^6 Hz) give potentially 10^{14} operations, or bits per second per neuron.

But increased information processing alone does not solve all problems regarding consciousness in the brain. Penrose Hameroff Orch OR further proposes tubulins can be quantum bits, or "qubits" in microtubule quantum computers, and that such quantum computations connect conscious brain functions to the most basic level of the universe.

This opens the door to consciousness being nonlocal, and in some cases possibly untethered to body and brain. These speculations are based on ideas in physics put forth by Sir Roger Penrose. It should be stated clearly that Sir Roger does not necessarily endorse the further speculations developed here, and generally avoids connections between science, religion, and spirituality.

Penrose defined OR self-collapse of superpositions (due to separations in space-time geometry) and moments of consciousness by $E = \hbar/t$. E is the gravitational self-energy of an object (or its equivalent space-time geometry) separated from itself. $\hbar$ is Planck's constant (over 2π) and t is the time at which OR occurs. E may be calculated based on factors including (1) the object's mass, (2) the level at which the object separates from itself, i.e., its entire mass, individual atoms, atomic nuclei, or subatomic particles, and (3) the spatial separation distance, how far the object, or its space-time geometry separates from itself. If a superposition of self-energy E evolves and avoids decoherence to reach time t, an OR moment of consciousness occurs.

Because of the inverse relation, the larger the mass and spatial separation E, the briefer the time t at which OR conscious moments occur. Superpositions E must avoid decoherence (i.e., the quantum system must be isolated from the classical environment) until time t is reached. Hence, the conditions for Penrose OR and conscious moments are fairly stringent.

Penrose and Hameroff suggest such conditions have evolved in the brain, specifically in microtubules inside brain neurons, and that microtubules perform quantum computations which are "orchestrated" by synaptic inputs and neurophysiology, isolated from decoherence, and terminated by Penrose OR, hence orchestrated objective reduction, "Orch OR." Microtubule quantum superpositions E are proposed to extend and entangle from neuron to neuron through gap junctions (which mediate gamma synchrony), enabling selective brain-wide quantum coherence among microtubules. Decoherence is suggested to be avoided through coherent pumping, actin gelation, ordered water and topological resonances. OR events also entail backward time effects, consistent with evidence for backward referral of conscious experience in the brain (Libet 1979). Entanglement with the future may enable real-time conscious action, and rescue consciousness from the unfortunate role of epiphenomenal illusion (Hameroff 2007).

Orch OR has been criticized since its inception in 1995, mainly because laboratory-built technological quantum computers require extreme cold to avoid decoherence by thermal vibrations, and the brain operates at warm biological temperatures (e.g., Tegmark 2000; Hagan et al. 2001). However, in the past 5 years numerous experiments have shown warm temperature quantum coherence in proteins involved in photosynthesis, ion channels, and other biomolecules (Engel et al. 2007). Dr. Anirban Bandyopadhyay (2010) at the National Institute of Materials Sciences in Tsukuba, Japan has preliminary evidence for quantum coherence, topological quantum conductance, and decoherence times of one-tenth millisecond or longer in individual microtubules at warm temperatures. For Orch OR and quantum biology, the future is fairly bright. Can Orch OR account for NDEs/OBEs and conceivably an after-life?

5.4 Orch OR, NDEs, and Altered States

Orch OR assumes consciousness typically occurs in the human brain at around 40 Hz, i.e., 40 conscious moments per second, corresponding with gamma synchrony EEG, the best measurable correlate of consciousness. For $t = 25$ ms (1/40 s), by

$E = \hbar/t$, E corresponds with nanograms of superpositioned tubulins ($\sim 10^{11}$ tubulins) distributed in microtubules in thousands of gap junction-connected neurons (and glia), still a very, very small fraction of the brain (total $\sim 10^{20}$ tubulins, 100 billion neurons).

In principle, OR and Orch OR (and thus conscious moments) can occur at any scale, in any type of medium, as long as superpositions avoid decoherence. Thus $E = \hbar/t$ predicts a full spectrum of possible conscious moments, much like the electromagnetic spectrum for photons. Large superpositions E will reach threshold quickly (and have more intense experiences) while small superpositions E will require longer times and have weak experiences (intensity proportional to E). For example, a single superpositioned electron (small E, long t), if isolated from environmental decoherence would reach threshold only after ten million years, and have an extremely low intensity moment of consciousness. Larger superpositions (large E, small t) will reach threshold quickly and have higher intensity consciousness. But decoherence must be avoided until time t and OR occurs. Higher levels of consciousness would involve larger E (more tubulins, more neurons and a higher portion of the brain), and shorter t, thus higher frequencies.

Vedic meditation, contemplation, and self-reflection exploring consciousness has led to descriptions of expanded states of consciousness or enlightenment involving 14 different levels, "astral planes," or "lokas." Lokas are portrayed as distinct worlds, realms, or planes of existence which differ from the 3-dimensional world of our everyday waking experience. Vedic texts say each plane or experienced reality has a characteristic frequency range, and is accessed or reached when it is matched by the frequency of the subject's conscious awareness (Chopra 2001).

Tibetan monks reach 80 Hz gamma synchrony during meditation (Lutz et al. 2004) presumably an enhanced, altered state, with twice as many conscious events per second, each at higher intensity. Magneto-encephalography has recorded coherent signals in the range of a kilohertz (1,000 Hz) from human brain (Papadelis et al. 2009), and higher frequency effects (megahertz, gigahertz, terahertz) have been measured in microtubules inside neurons (Bandyopadhyay 2010). Could consciousness shift levels to higher frequencies and greater brain involvement in altered and enhanced states?

Electrical signals occur in the brain in a self-similar way at different spatial and temporal scales, scale-free dynamics (He et al. 2010). This is also called pink noise, proportional to $1/f^{\alpha}$, where f is frequency and α is the separations in scale (e.g., spatial and temporal orders of magnitude) at which the information repeats, similar to a fractal or hologram.

Fractal or holographic-like structure also occurs in "small world" and "large world" networks of neurons, nested hierarchies of networks within networks within networks. And inside neurons are cytoskeletal networks including microtubules which may also process information. Scale-free dynamics occurs both temporally and structurally in the brain, in layers or information processing systems with both bottom–up and top–down relationships.

In altered states, the process of consciousness may shift to different planes, or scales in the brain, with higher frequencies (smaller t), greater intensity, and larger

E in terms of number of involved microtubules, neurons, and volume of brain capacity. Consciousness occurring by $E = h/t$ normally at 40 Hz (each conscious moment involving roughly one millionth of brain microtubules) could transition to higher frequencies at, say 10 kHz, megahertz, gigahertz, and terahertz levels. These would involve greater and greater proportions of brain neurons and microtubules. These levels would involve, respectively, 1/10,000th, 1/100th, and, for gigahertz consciousness, the entire brain.

Thus altered states of consciousness may involve transcendence to deeper, more intense levels of experience, deeper levels of reality, e.g., consistent with Vedic astral planes or lokas, and enlightenment reached by meditation and spiritual practices. Such enhanced, altered states need not involve alternative dimensions or universes, but rather deeper, more fine scale geometry in nonlocal holographic-like levels or scales in this one universe.

As the Beatles said (Lennon and McCartney 1968): "The deeper you go, the higher you fly, The higher you fly, the deeper you go."

At any frequency, Orch OR consciousness in the brain is occurring in fundamental space-time geometry, localized to brain neuronal microtubules and driven by metabolic processes. When the blood stops flowing, energy and oxygen depleted and microtubules inactivated or destroyed (e.g., NDE/OBE, death), it is conceivable that the quantum information which constitutes consciousness could shift to deeper planes and continue to exist purely in space-time geometry, outside the brain, distributed nonlocally. Movement of consciousness to deeper planes could account for NDEs/OBEs, as well as, conceivably, a soul apart from the body.

5.5   End-of-Life Brain Activity

Gamma synchrony EEG brain activity is known to correlate with normal consciousness. Monitors able to measure and detect gamma synchrony and other correlates of consciousness have been developed for use during anesthesia to provide an indicator of depth of anesthesia and prevent intraoperative awareness, i.e., to avoid patients being conscious when they are supposed to be anesthetized and unconscious. For example the "BIS" monitor (Aspect Medical Systems, Newton MA) records and processes frontal EEG to produce a digital "bispectral index," or BIS number on a scale of 0–100. A BIS number of 0 equals EEG silence, and 80–100 is the expected value in a fully awake, conscious adult with gamma synchrony. A BIS number maintained between 40 and 60 is recommended for general anesthesia. The "SEDline" monitor (Hospira, Lake Forest, IL) also records frontal EEG and produces a comparable 0–100 index.

In recent years, these monitors have been applied outside of anesthesiology, e.g., to dying patients at or near the moment of death, revealing startling end-of-life brain activity.

In a study reported in the *Journal of Palliative Medicine*, Chawla et al. (2009) reported on seven critically ill patients from whom life support (medications, machine ventilation) was being withdrawn, allowing them to die peacefully. As per protocol, they were monitored with a BIS or SEDline brain monitor during the process of dying. While on life support the patients were neurologically intact but heavily sedated, with BIS or SEDline numbers near 40 or higher. Following withdrawal, the BIS/SEDline generally decreased to below 20 after several minutes, at about the time cardiac death occurred. This was marked by lack of measurable arterial blood pressure or functional heartbeat. Then, in all seven patients postcardiac death, there was a burst of activity as indicated by abrupt rise of the BIS or SEDline to between 60 and (in most cases) 80 or higher. After a period of such activity ranging from 90 s to 20 min, the activity dropped abruptly to near zero.

The SEDline number is derived from a proprietary algorithm which includes EEG data. In one patient, raw SEDline data was analyzed and revealed the burst of postcardiac death brain activity to include gamma synchrony, an indicator of conscious awareness. Chawla et al. raise the possibility that the measured postcardiac death brain activity might correlate with NDEs/OBEs. Of course the patients died, so we have no confirmation that such experiences occurred.

In another study published in the journal *Anesthesia and Analgesia*, Auyong et al. (2010) described three brain-injured patients from whom medical and ventilatory support were withdrawn prior to "postcardiac death" organ donation (Csete 2010) These patients were hopelessly brain-damaged, but technically not brain dead. Their families consented to withdrawal of support and organ donation. Such patients are allowed to die "naturally" after withdrawal of support, their bodies then quickly taken to surgery for organ donation.

The three patients in the Auyong et al. study prior to withdrawal of support had BIS numbers of 40 or lower, with one near zero. Soon after withdrawal, near the time of cardiac death, the BIS number dwindled downward and then spiked to approximately 80 in all three cases, and remained there for 30–90 s. The number then abruptly returned to near zero, followed thereafter by declaration of death and organ donation. Various sources of artifact for the end-of-life brain activity were considered and excluded.

Obviously we cannot say whether the end-of-life brain activity is indeed related to NDEs/OBEs, or even possibly the soul leaving the body. Nor do we know how commonly it occurs (ten out of ten in the two studies cited). Those issues aside, the mystery remains as to how brain activity occurs in metabolically dead tissue, receiving no blood flow or oxygen, and lacking mechanisms to remove toxic metabolites.

Some describe the end-of-life brain activity as nonfunctional, generalized neuronal depolarization. Chawla et al. suggested excess extracellular potassium could cause "last gasp" neuronal spasms of activity throughout the brain. Another suggested cause is calcium-induced programmed neuronal death by apoptosis. But such explanations seem unable to account for globally organized coherent synchrony during the end-of-life brain activity.

If end-of-life brain activity does correlate with conscious NDE/OBE phenomenology and/or the soul leaving the body, we still face the question of how/why conscious activity, or even synchronized activity of any sort is occurring in the nearly dead brain. But there are logical possibilities.

Energy requirements for consciousness may be small compared to nonconscious brain functions, especially if consciousness occurs primarily in dendrites and cell bodies rather than axonal firings. Neuronal hypoxia and acidosis would disable sodium-potassium ATPase pumps, preventing axonal action potentials, but temporarily sparing lower energy dendritic activity. Consciousness as a low energy quantum process might transiently flourish if energy-dependent decoherence-causing mechanisms were impaired, resulting in a transient burst of enhanced consciousness.

In the Orch OR context, consciousness occurs as a process at the level of fundamental space-time geometry. When the brain is under duress, it is conceivable quantum information processes constituting consciousness dissipate to the nonlocal universe at large. A dualist perspective, in which a separate, as yet undefined spiritual information field constitutes awareness outside the body, may not be necessary. An afterlife, an actual soul-as-quantum information leaving the body and persisting as entangled fluctuations in multiple scales, or planes in quantum space-time geometry, may be scientifically possible.

5.6   Conclusion: The Quantum Soul

Conventional science and philosophy attempt to base consciousness strictly on classical physics, rejecting the possibility of quantum nonlocality in consciousness, including persistence outside the body as indicated by NDEs/OBEs, religious lore, and anecdotal memories suggesting reincarnation. But evidence in recent years links biological functions to quantum processes, raising the likelihood that consciousness depends on nonlocal quantum effects in the brain. That in turn suggests that the "hard problem" of the nature of conscious experience requires a worldview in which consciousness or its precursors are irreducible components of reality, fundamental space-time geometry at the Planck scale. Max Planck (1931) was clear-sighted when he said, "I regard consciousness as fundamental. We cannot get behind consciousness." Vedic and other spiritual traditions have similar assumptions; consciousness and knowledge are intrinsic to the universe.

How did they get there? Physicist Paola Zizzi has proposed that the period of rapid inflation during the very early Big Bang was characterized by superposition of multiple possible universes. By $E = \hbar/t$, Zizzi (2004) has calculated that the end of inflation and selection of this universe was caused by a cosmic conscious moment at a particular instant during the Big Bang (the "Big Wow"). Perhaps the possible universes were related to a previous universe, as Penrose (2010) has proposed in "Cycles of Time," our universe mutating and evolving with each rebirth[1].

The Penrose-Hameroff Orch OR model of consciousness proposes a connection between quantum brain processes and fundamental space-time geometry. In this study we consider Orch OR in the context of anecdotal reports of NDE/OBE experiences as well as circumstantial evidence for afterlife, reincarnation, and the potential

---

[1] Although this proposal does not include inflation.

for quantum consciousness in space-time geometry. We conclude the concept of a "quantum soul" is scientifically plausible.

The "quantum soul" implies consciousness in the brain as described by Orch OR, as well as nonlocal features including:

1. Interconnectedness via entanglement among living beings and the universe
2. Contact with cosmic wisdom/Platonic values embedded as quantum information in fundamental space-time geometry
3. Consciousness as patterns in nonlocal fractal/holographic-like space-time geometry, able to exist at deeper planes and scales independent of biology

We present a secular, scientific approach consistent with all religions and known science. With the advent of quantum biology, nonlocality in consciousness must be taken seriously, potentially building a bridge between science and spirituality.

References

Auyong, D. B., Klein, S. M., Gan, T. J., Roche, A. M., Olson, D. W., & Habib, A. S. (2010). Processed electroencephalogram during donation after cardiac death. *Anesthesia and Analgesia, 110*(5), 1428–1432.

Bandyopadhyay, A. (2010). *Direct experimental evidence for quantum states in microtubules and topological invariance.* Toward a Science of Consciousness 2011 Abstracts. Retrieved, from http://www.consciousness.arizona.edu (manuscript in preparation).

Blackmore, S. J. (1998). Experiences of anoxia: Do reflex anoxic seizures resemble near-death experiences? *Journal of Near Death Studies, 17,* 111–120.

Blanke, O., Landis, T., Spinelli, L., & Seeck, M. (2004). Out-of-body experience and autoscopy of neurological origin. *Brain, 127*(2), 243–258.

Brunden, K. R., Yao, Y., Potuzak, J. S., Ferrer, N. I., Ballatore, C., James, M. J., et al. (2011). The characterization of microtubule-stabilizing drugs as possible therapeutic agents for Alzheimer's disease and related tauopathies. *Pharmacological Research, 63*(4), 341–351.

Chalmers, D. J. (1996). *The conscious mind – in search of a fundamental theory.* New York: Oxford University Press.

Chawla, L. S., Akst, S., Junker, C., Jacobes, B., & Seneff, M. G. (2009). Surges of electroencephalogram activity at the time of death: A case study. *Journal of Palliative Medicine, 12*(12), 1095–1100.

Chopra, D. (2001). *How to know god: The soul's journey into the mystery of mysteries* New york, NY. Running Press Book Publishers.

Chopra, D. (2006). *Life after death – the burden of proof.* New York: Three Rivers Press.

Chown, M. (2009). Our world may be a giant hologram. *NewScientist.* Retrieved, from http://www.newscientist.com/article/mg20126911.300.2010-04-19

Christie, J. M., & Westbrook, G. L. (2006). Lateral excitation within the olfactory bulb. *Journal of Neuroscience., 26*(8), 2269–2277.

Crick, F. C., & Koch, C. (2001). A framework for consciousness. *Nature Neuroscience, 6,* 119–126.

Csete, M. (2010). Donation after cardiac death and the anesthesiologist. *Anesthesia and Analgesia, 5,* 1253–1254.

Davies, P. (2006). *The Goldilocks enigma.* London: Allen Lane.

Dennett, D. C. (1991). *Consciousness explained.* Boston: Little, Brown.

Dermietzel, R. (1998). Gap junction wiring: A 'new' principle in cell-to-cell communication in the nervous system? *Brain Research Reviews, 26*(2–3), 176–183.

Engel, G. S., Calhoun, T. R., Read, E. L., Ahn, T.-K., Mancal, T., Cheng, Y.-C., et al. (2007). Evidence for wavelike energy transfer through quantum coherence in photosynthetic systems. *Nature, 446,* 782–786.

Gray, C. M., & Singer, W. (1989a). Stimulus-specific neuronal oscillations in orientation columns of cat visual cortex. *Proceedings of the National Academy of Sciences USA* (Vol. 86, pp. 1698–1702). USA.

Gray, C. M., & Singer, W. (1989b). Stimulus-specific neuronal oscillations in orientation columns of cat visual cortex. *Proceedings of the National Academy of Sciences USA* (Vol. 86, pp. 1698–1702). USA.

Greyson, B. (1993). Varieties of near-death experience. *Psychiatry, 56*(4), 390–399.

Gurzadyan, V. G., & Penrose, R. (2010). *Concentric circles in WMAP data may provide evidence of violent pre-Big-Bang activity.* arXiv:1011.3706.

Hameroff, S. (1998a). Quantum computation in brain microtubules? The Penrose Hameroff "Orch OR" model of consciousness. *Philosophical Transactions of the Royal Society of London Series A, 356,* 1869–1896.

Hameroff, S. (1998b). Quantum computation in brain microtubules – the Penrose-Hameroff "Orch OR" model of consciousness. *Philosophical Transactions of the Royal Society of London Series A, 356,* 1869–1896.

Hameroff, S. (2006). Consciousness, neurobiology and quantum mechanics: The case for a connection. In Tuszyuski J (Ed.), *The emerging physics of consciousness.* New York: Springer.

Hameroff, S. (2007). The brain is both neurocomputer and quantum computer. *Cognitive Science, 31,* 1035–1045.

Hameroff, S. (2010). The "conscious pilot"-dendritic synchrony moves through the brain to mediate consciousness. *Journal of Biological Physics, 36*(1), 71–93.

Hameroff, S. R., & Penrose, R. (1996a). Orchestrated reduction of quantum coherence in brain microtubules: A model for consciousness. In S. R. Hameroff, A. Kaszniak, & A. C. Scott (Eds.), *Toward a science of consciousness the first Tucson discussions and debates* (pp. 507–540). Cambridge: MIT Press. Also published in Mathematics and Computers in Simulation (1996) 40:453–480.

Hameroff, S. R., & Penrose, R. (1996b). Conscious events as orchestrated spacetimeselections. *Journal of Consciousness Studies, 3*(1), 36–53.

Hawking, S., & Mlodinow, L. (2010). *Grand design.* New York: Bantam.

Hebb, D. O. (1949). *Organization of behavior: A neuropsychological theory.* New Yourk: Wiley.

Hodgson, D. (2007). Making our own luck. *Ratio, 20,* 278–292.

Hogan, C. J. (2008). Measurement of quantum fluctuations in geometry. *Physical Review D, 77*(10), 104031. doi: 10.1103/PhysRevD.77.104031.arXiv:0712.3419

Huxley, T. H. (1893). Method and results: Essays.

Jansen, K. L. (2000). A review of the nonmedical use of ketamine: Use, users and consequences. *Journal of Psychoactive Drugs, 32*(4), 419–433.

Koch, C. (2004). *The quest for consciousness: A neurobiological approach.* Englewood: Roberts and Company.

Koch, C., & Crick, F. (2001). The zombie within. *Nature, 411,* 893.

Lennon, J., & McCartney, P. (1968). *Everybody's got something to hide except for me and my monkey.* White Album. Sony /ATV Music, Nashville,TN.

Lutz, A., Greischar, L. L., Rawlings, N. B., Ricard, M., & Davidson, R. J. (2004). Long-term meditators self-induce high-amplitude gamma synchrony during mental practice. *The Proceedings of the National Academy of Sciences USA, 101*(46), 16369–16373.

Nadeau, R., & Kafatos, M. (2001). *The non-local universe: The new physics and matters of the mind.* Oxford: Oxford University Press.

Papadelis, C., Poghosyan, V., Fenwick, P. B., & Ioannides, A. A. (2009). MEG's ability to localise accurately weak transiently neural sources. *Clinical Neurophysiology, 120*(11), 1958–1970.

Parnia, S., Spearpoint, K., & Fenwick, P. B. (2007). Near death experiences, cognitive function and psychological outcomes of surviving cardiac arrest. *Resuscitation, 74*(2), 215–221.

Penrose, R. (2004). *The road to reality: A complete guide to the laws of the universe.* London: Vintage Books.

Penrose, R. (2010). *Cycles of time: An extraordinary new view of the universe.* London: The Bodley Head.

Penrose, R., & Hameroff, S. R. (1995). Gaps, what gaps? Reply to Grush and Churchland. *Journal of Consciousness Studies, 2*(2), 99–112.

Planck, M. (1931). The observer, London, Januvary 29, 1931.

Singer, W. (1999). Neuronal synchrony: A versatile code for the definition of relations. *Neuron, 24*, 111–125.

Singer, W., & Gray, C. M. (1995). Visual feature integration and the temporal correlation hypothesis. *Annual Review of Neuroscience, 18*, 555–586.

Susskind, L. (1994). *The world as a hologram.* Retrieved, from http://arxiv.org/abs/hep-th/9409089

van Lommel, P., van Wees, R., Meyers, V., & Elfferich, I. (2001). Near-death experience in survivors of cardiac arrest: A prospective study in The Netherlands. *Lancet, 358*(9298), 2039–2045.

Wegner, D. M. (2002). *The illusion of conscious will.* Cambridge: MIT Press.

Zizzi, P. A. (2004). *Emergent consciousness: From the early universe to our mind.* Retrieved, from http://arxiv.org/abs/gr-qc/0007006

Part III
Functional Neuroimaging

Chapter 6
The Neurobiological Correlates of Meditation and Mindfulness

Jesse Edwards, Julio Peres, Daniel A. Monti, and Andrew B. Newberg

Abstract Mindfulness refers to a calm awareness of cognitions, sensations, emotions, and experiences. This state is frequently achieved through mindfulness meditation (MM) which is a practice that cultivates non-judgmental awareness of the present moment. MM has also become widely used in a variety of psychological, medical, and wellness populations. Recently, there have been a number of studies that have elucidated some of the neurophysiological processes involved with MM and other similar meditation practices. This chapter provides a review of that literature, which includes neuroanatomy, neurophysiology, neurotransmitter systems, and recent brain-imaging advances.

6.1 Introduction

Meditation has been widely studied due to its potential health benefits (Baer 2003) and its status as a special state of consciousness, which is uniquely different from both ordinary wake and sleep states (Baerentsen et al. 2010). Several results suggest that meditation techniques longitudinally reduce the affective/motivational dimension of the brain's response to pain (Kakigi et al. 2005; Orme-Johnson et al. 2006). In fact, individuals' perceptions are decisive for the quality of their interaction with the environment (Peres et al. 2005), so the experimental study of meditation – and related state of consciousness, perceptual representations of the world and behaviors – is of extreme interest within the scope of mental health research.

J. Edwards • D.A. Monti • A.B. Newberg (✉)
Myrna Brind Center of Integrative Medicine, Thomas Jefferson University and Hospital,
Philadelphia, PA 19107, USA
e-mail: Andrew.newberg@jefferson.edu

J. Peres
ProSer – Institute of Psychiatry, Universidade de São Paulo, São Paulo, Brazil

A. Moreira-Almeida and F.S. Santos (eds.), *Exploring Frontiers of the Mind-Brain Relationship*, Mindfulness in Behavioral Health, DOI 10.1007/978-1-4614-0647-1_6,
© Springer Science+Business Media, LLC 2012

Perception may be seen as a process of deconstruction and reconstruction of the external world on the basis of the patterns of stimulations that excite our sensory receptors (Palmer 1999). The underlying principle in perception mechanisms involves extracting statistical correlations fimpressions from the world in order to build models that are temporarily useful for adaptive interrelation with the environment (Ramachandran and Gregory 1991; Ramachandran et al. 1998). Once the state of consciousness shifts, perception of the event changes too (Dietrich 2003) and, consequently, there is a new interaction and relationship with the surrounding (Peres et al. 2005, 2007).

Thus, the neurobiological study of mindfulness and meditation techniques may prove to be a highly important area of research in the next decade. This area of inquiry can offer a fascinating window into human consciousness, psychology, and experience; the relationship between mental states and physiology; emotional and cognitive processing; and the biological correlates of religious experience.

Mindfulness meditation (MM) engages the mind by intentional reflection upon concepts such as awareness, acceptance, release, attention, and nonjudgment (Ivanovski and Malhi 2007). A principal objective of such reflection practices is to alter one's relationship to his or her thoughts – not to attempt to change or control the thoughts themselves. Meditators embrace a dialectical state of "willful passivity" in their mind and their relationship to environment. Thus, meditation may change the perception and corresponding behaviors.

In the past 30 years, scientists have been able to explore the biological effects and mechanism of meditation in much greater detail, largely due to the development of more advanced imaging technologies. Initial studies measured changes in autonomic activity, such as heart rate and blood pressure, as well as electroencephalographic changes. More recent studies have explored changes in hormonal and immunological functions associated with meditation. Studies also have explored the clinical effects of meditation in both physical and psychological disorders. Generally, these studies have methodological limitations due to sample size, the ability to delineate the type of meditation, and ultimately the ability to identify the meditation practice as the mediating factor in physical or mental health benefits. However, these studies do suggest that MM can have a positive impact on a range of psychological symptoms such as anxiety, depression, and various mood disorders, as well as stress-related physical symptoms such as pain and fatigue (Monti et al. 2006).

Functional neuroimaging has opened a new window into the investigation of meditative states by exploring the neurological correlates of these experiences. A growing number of neuroimaging studies of mindfulness and other meditation practices are currently available in the literature. The neuroimaging techniques include positron emission tomography (PET; Herzog et al. 1990–1991; Lou et al. 1999), single photon emission computed tomography (SPECT; Newberg et al. 2001), and functional magnetic resonance imaging (fMRI; Lazar et al. 2000). Each of these techniques provides different advantages and disadvantages in the study of meditation. Functional MRI, while having improved resolution over SPECT and the ability of immediate anatomic correlation, would be very difficult to utilize for the study of meditation because of noise from the machine and the problem of requiring the subject to lie down, an atypical posture for many forms of meditation. In fact, we attempted the use of fMRI with one of our initial meditation subjects in order to determine feasibility, but the subject found it extremely difficult to carry out the meditation practice. While PET imaging also provides better resolution than SPECT, if one strives to make the environment relatively distraction free to maximize the chances of having a strong meditative experience, it is sometimes beneficial to perform these studies during nonclinical times, which may make PET radiopharmaceuticals such as fluorodeoxyglucose difficult to obtain. Thus, functional brain imaging offers important techniques for studying meditation, although the best approach may depend on a number of individual factors.

In this chapter, we review the existing data on neurophysiology and physiology with relation to meditation practices. It is noted that there are many possible neurochemical changes that may occur during meditation, even though they may not occur in every type of practice or in each individual. MM itself is a practice which varies significantly in form, purpose, and impact. Although studies of MM vary in their quality, target populations, and outcome measures, there are numerous commonalities amongst the results and, consequently, the implications of their respective findings.

6.2 Meditation

Meditation is an ancient spiritual practice, which aims to still the mind by reducing negative thoughts, inducing a state of relaxed attention, and increasing feelings of. Although there are many different schools and types of meditation – connected to formal philosophical systems and/or religious practices (e.g., Hindu, Buddhist, Muslim, Christian, etc.) – the various forms and practices can be seen as variations of concrete operationalizations of meditation. Therefore, when compared to other states and processes of the mind, these differences may be disregarded as inessential in comparison with the common nucleus that characterises meditation in general (Baerentsen et al. 2010). In terms of the larger topic of meditation, in addition to MM, the most common other type involves purposeful attention on a particular object, image, phrase, or word. This form of meditation is designed to lead to a subjective experience of absorption with the object of focus – a dissolution of the differentiation of self and object. There is another distinction in which meditation is guided by following along with a leader who verbally directs the practitioner, either in person or on tape. Others merely practice the meditation on their own volition. We might expect that this difference between volitional and guided meditation should also be reflected in specific differences in cerebral activation.

Phenomenological analysis suggests that the end result of many practices of meditation is similar, although this result might be described using different characteristics depending on the culture and individual. Therefore, it seems reasonable that while the initial neurophysiological activation occurring during any given practice may differ, there should eventually be a convergence of data.

6.3 Activation of the Prefrontal and Cingulate Cortex

Brain imaging studies suggest that willful acts and tasks that require sustained attention are initiated via activity in the prefrontal cortex (PFC), particularly in the right hemisphere (Frith et al. 1991; Pardo et al. 1991; Ingvar 1994). The cingulate gyrus has also been shown to be involved in focusing attention, probably in conjunction with the PFC (Vogt et al. 1992). Since meditation requires intense focus of attention, it seems appropriate that a model for meditation begin with activation of the PFC (particularly the right), as well as the cingulate gyrus. This notion is supported by the increased activity observed in these regions on several of the brain imaging studies of volitional types of meditation. Therefore, many meditation practices, particularly MM, appear to start by activating the prefrontal and cingulate cortex associated with the will or intent to clear the mind of thoughts or to focus on an object. Thus, prefrontal and cingulate activation may be associated with the volitional and attentional aspects of meditation.

6.4 Thalamic Activation

Several animal studies have shown that the PFC, when activated, innervates the reticular nucleus of the thalamus (Cornwall and Phillipson 1988), particularly as part of a more global attentional network (Portas et al. 1998) (Fig. 6.1).

Such activation may be accomplished by the PFC's production and distribution of the excitatory neurotransmitter glutamate which the PFC neurons use to

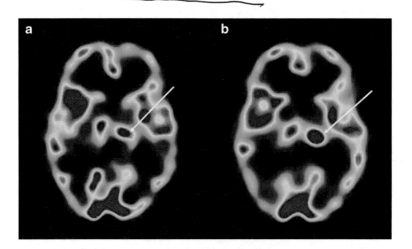

Fig. 6.1 Two SPECT scans showing cerebral blood flow changes in the thalamus (*arrows*) between the resting state (**a**) and the meditation state (**b**). The scans show a significant increase in the left thalamus (on *right side* of figure) as demonstrated by increasing color of red on the scan. This finding supports the notion that the thalamus is important in the meditation practice

communicate among themselves and to innervate other brain structures (Cheramy et al. 1987). The thalamus itself governs the flow of sensory information to cortical processing areas via its interactions with the lateral geniculate and lateral posterior nuclei and also likely uses the glutamate system in order to activate neurons in other structures (Armony and LeDoux 2000). The lateral geniculate nucleus receives raw visual data from the optic tract and routes it to the striate cortex for processing (Andrews et al. 1997). The lateral posterior nucleus of the thalamus provides the posterior superior parietal lobule (PSPL) with the sensory information it needs to determine the body's spatial orientation (Bucci et al. 1999).

When excited, the reticular nucleus secretes the inhibitory neurotransmitter gamma aminobutyric acid (GABA) onto the lateral posterior and geniculate nuclei, cutting off input to the PSPL and visual centers in proportion to the reticular activation (Destexhe et al. 1998). During meditation, because of the increased activity in the PFC, particularly on the right, there should be a concomitant increase in the activity in the reticular nucleus of the thalamus. While brain imaging studies of meditation have not had the resolution to distinguish the reticular nuclei, our SPECT study of Tibetan Buddhist meditation did demonstrate a general increase in thalamic activity that was proportional to the activity levels in the PFC (Newberg et al. 2001). This is consistent with, but does not confirm the specific interaction between the PFC and reticular nuclei. If the activation of the right PFC causes increased activity in the reticular nucleus during meditation, the result may be decreased sensory input entering into the PSPL. Several studies have demonstrated an increase in serum GABA during meditation, possibly reflecting increased central GABA activity (Elias et al. 2000). This functional deafferentation related to increased GABA would mean that fewer distracting outside stimuli would arrive at the visual cortex and PSPL enhancing the sense of focus. For example, a study of novice and expert meditators evaluated fMRI during meditation sessions while external distractions were induced, in an effort to assess each practitioner's degree of involuntary response to the stimuli (Brefczynaki-Lewis et al. 2007). In each of the sessions, meditation reduced (to varying degrees, mostly correlating to the meditator's level of experience) practitioners' susceptibility to uncontrolled response to distracting stimuli. Another study revealed that compassion meditation enhances the emotional and somatosensory brain representations of others' emotions, and that this effect is modulated by level of expertise of the practitioner (Lutz et al. 2009).

It should also be noted that the dopaminergic system, via the basal ganglia, is believed to participate in regulating the glutamatergic system and the interactions between the PFC and subcortical structures. A PET study utilizing 11C-Raclopride to measure the dopaminergic tone during Yoga Nidra meditation demonstrated a significant increase in dopamine levels during the meditation practice (Kjaer et al. 2002). They hypothesized that this increase may be associated with the gating of cortical–subcortical interactions that leads to an overall decrease in readiness for action that is associated with this particular type of meditation. Future studies will be necessary to elaborate on the role of dopamine during meditative practices as well as the interactions between dopamine and other neurotransmitter systems.

6.5 Parietal Lobe Deafferentation

The PSPL is heavily involved in the analysis and integration of higher-order visual, auditory, and somaesthetic information (Adair et al. 1995). It is also involved in a complex attentional network that includes the PFC and thalamus (Fernandez-Duque and Posner 2001). Through the reception of auditory and visual input from the thalamus, the PSPL is able to help generate a three-dimensional image of the body in space, provide a sense of spatial coordinates in which the body is oriented, help distinguish between objects, and exert influences in regard to objects that may be directly grasped and manipulated (Mountcastle et al. 1980; Lynch 1980). These functions of the PSPL might be critical for distinguishing between the self and the external world. It should be noted that a recent study has suggested that the superior temporal lobe may play a more important role in body spatial representation, although this has not been confirmed by other reports (Karnath et al. 2001). However, it remains to be seen what is the actual relationship between the parietal and temporal lobes in terms of spatial representation.

Regardless, deafferentation of these orienting areas of the brain, we have previously proposed, is an important concept in the physiology of meditation (Newberg and Iversen 2003). If, for example, deafferentation of the PSPL by the reticular nucleus's GABAergic effects, the person may begin to lose their usual ability to spatially define the self and help to orient the self. Alterations in the sense or awareness of the self or the manner in which the self orients toward the rest of the world are often reported in MM. Such a notion is supported by clinical findings in patients with parietal lobe damage who have difficulty orienting themselves, a well-known condition called Balint syndrome. The effects of meditation are likely to be more selective and do not destroy the sense of self, but alter the perception of it. Deafferentation of the PSPL has also been supported by SPECT imaging studies demonstrating decreased activity in this region during intense meditation (Newberg et al. 2001, 2003).

6.6 Hippocampal and Amygdalar Activation

In addition to the complex cortical-thalamic activity, MM might also be expected to alter activity in the limbic system since mindfulness attempts to enhance awareness of emotional responses. It has also been reported that stimulation of limbic structures, particularly the hippocampus, is associated with experiences similar to those described during various meditation states (Fish et al. 1993; Saver and Rabin 1997). The hippocampus acts to modulate and moderate cortical arousal and responsiveness, via rich and extensive interconnections with the PFC, other neocortical areas, the amygdala, and the hypothalamus (Joseph 1996). Hippocampal stimulation has been shown to diminish cortical responsiveness and arousal; however, if cortical arousal is initially at a low level, then hippocampal stimulation tends to augment cortical activity (Redding 1967). The ability of the hippocampus to stimulate or inhibit neuronal activity in other structures likely relies upon the glutamate and GABA systems, respectively. The partial deafferentation of the right PSPL during meditation may result in the stimulation of the right hippocampus because of the inverse modulation of the hippocampus in relation to cortical activity. If, in addition, there is simultaneous direct stimulation of the right hippocampus via the thalamus (as part of the known attentional network) and mediated by glutamate, then a powerful recruitment of stimulation of the right hippocampus occurs. Right hippocampal activity may ultimately enhance the stimulatory function of the PFC on the thalamus via the nucleus accumbens, which gates the neural input from the PFC to the thalamus via the neuromodulatory effects of dopamine (Newman and Grace 1999; Chow and Cummings 1999).

The hippocampus greatly influences the amygdala, such that they complement and interact in the generation of attention, emotion, and certain types of imagery. It seems that much of the prefrontal modulation of emotion is via the hippocampus and its connections with the amygdala (Poletti and Sujatanond 1980). Because of this reciprocal interaction between the amygdala and hippocampus, the activation of the right hippocampus may stimulate the right lateral amygdala if the individual experiences heightened emotional responses. The results of the fMRI study by Lazar et al. (2000) also support the notion of increased activity in the regions of the amygdala and hippocampus during meditation. On the other hand, studies of MM in particular have reported enhanced PFC activity in conjunction with decreased activity in the amygdala, which corresponds with diminished reactivity to emotional stimuli (Lieberman et al. 2007; Creswell et al. 2007). Thus, different types of meditation practices may result in different activity levels in the limbic structures depending on whether emotional responses are enhanced or diminished.

6.7 Hypothalamic and Autonomic Nervous System (ANS) Changes

Activity in the right lateral amygdala has been shown to modulate activity in the ventromedial portion of the hypothalamus, which can result in either excitation or stimulation of the peripheral parasympathetic system (Davis 1992). Increased parasympathetic activity should be associated with the subjective sensation first of relaxation, and eventually, of a more profound quiescence. Activation of the parasympathetic system would also cause a reduction in heart rate and respiratory rate. All of these physiological responses have been observed during meditation (Jevning et al. 1992). In accord with the Indo-Tibetan tradition of self-healing, one study narrowed its analysis of MM specifically to that of mindfulness-based stress reduction; meditators experienced a notable reduction of stress levels, along with the secretion of hormones (such as cortisol) associated with stress response (Peterson et al. 2009).

Typically, when breathing and heart rate slow down, the paragigantocellular nucleus of the medulla ceases to innervate the locus ceruleus (LC) of the pons. The LC produces and distributes norepinephrine (NE), (Foote 1987), a neuromodulator that increases the susceptibility of brain regions to sensory input by amplifying strong stimuli, while simultaneously gating out weaker activations and cellular "noise" that fall below the activation threshold (Waterhouse et al. 1998). Decreased stimulation of the LC results in a decrease in the level of NE (Van Bockstaele and Aston-Jones 1995). The breakdown products of catecholamines such as NE and epinephrine have generally been found to be reduced in the urine and plasma during meditation (Walton et al. 1995; Infante et al. 2001), which may simply reflect the systemic change in autonomic balance. However, it is not inconsistent with a cerebral decrease in NE levels as well. During a meditative practice, the reduced firing of the paragigantocellular nucleus probably cuts back its innervation of the LC, which densely and specifically supplies the PSPL and the lateral posterior nucleus with NE. Thus, a reduction in NE would decrease the impact of sensory input on the PSPL, contributing to its deafferentation. Similar to the study by Brefczynaki-Lewis et al. pertaining to meditators' distractibility, another study measured the influence of external stimuli on meditators by inducing selected pain stimuli (Orme-Johnson et al. 2006). Results indicated a decreased cerebral response to such stimuli, not only during the act of meditation, but even in the baseline state of long-term meditation practitioners. The clinical applications of findings pertaining to the brain's stimulus–response mechanisms are quite promising, particularly in the treatment of chronic ailments and some psychiatric disorders.

The LC would also deliver less NE to the hypothalamic paraventricular nucleus. The paraventricular nucleus of the hypothalamus typically secretes corticotropin-releasing hormone (CRH) in response to innervation by NE from the LC (Ziegler et al. 1999). This CRH stimulates the anterior pituitary to release adrenocorticotropic hormone (ACTH) (Livesey et al. 2000). ACTH, in turn, stimulates the adrenal cortex to produce cortisol, one of the body's stress hormones (Davies et al. 1985). Decreasing NE from the LC during meditation would likely decrease the production of CRH by the paraventricular nucleus and ultimately decrease cortisol levels. Most studies have found that urine and plasma cortisol levels are decreased during meditation (Sudsuang et al. 1991; Jevning et al. 1978).

The drop in blood pressure associated with parasympathetic activity during MM practices would be expected to relax the arterial baroreceptors leading the caudal ventral medulla to decrease its GABAergic inhibition of the supraoptic nucleus of the hypothalamus. This lack of inhibition can provoke the supraoptic nucleus to release the vasoconstrictor arginine vasopressin (AVP), thereby tightening the arteries and returning blood pressure to normal (Renaud 1996). AVP has also been shown to contribute to the general maintenance of positive affect (Pietrowsky et al. 1991), decrease self-perceived fatigue and arousal, and significantly improve the consolidation of new memories and learning (Weingartner et al. 1981). In fact, plasma AVP has been shown to increase dramatically during meditation (O'Halloran et al. 1985). The sharp increase in AVP should result in a decreased subjective feeling of fatigue and increased sense of arousal. It could also help to enhance the meditator's memory of his experience, perhaps explaining the subjective phenomenon that meditative experiences are remembered and described in very vivid terms. Given the above findings, it appears that meditation practices including mindfulness likely are associated with significant changes in the hypothalamus and ANS.

6.8   PFC Effects on Other Neurochemical Systems

As a MM practice continues, there should be continued activity in the PFC associated with the persistent will to focus on awareness. In general, as PFC activity increases, it produces ever-increasing levels of free synaptic glutamate in the brain. Increased glutamate can stimulate the hypothalamic arcuate nucleus to release beta-endorphin (BE) (Kiss et al. 1997). BE is an opioid produced primarily by the arcuate nucleus of the medial hypothalamus and distributed to the brain's subcortical areas (Yadid et al. 2000). BE is known to depress respiration, reduce fear, reduce pain, and produce sensations of joy and euphoria (Janal et al. 1984). Such effects have been described during meditation may implicate some degree of BE release related to the increased PFC activity. Meditation practices have been found to disrupt diurnal rhythms of BE and ACTH, while not affecting diurnal cortisol rhythms (Infante et al. 1998). However, it is likely that BE is not the sole mediator in such experiences during meditation because simply taking morphine-related substances does not produce equivalent experiences as in meditation. Furthermore, one very limited study demonstrated that blocking the opiate receptors with naloxone did not affect the experience or EEG associated with meditation (Sim and Tsoi 1992).

Glutamate activates N-methyl D-Aspartate receptors (NMDAr), but excess glutamate can kill these neurons through excitotoxic processes (Albin and Greenamyre 1992). We propose that if glutamate levels approach excitotoxic concentrations during intense states of meditation, the brain might limit its production of N-acetylated-alpha-linked-acidic dipeptidase, which converts the endogenous NMDAr antagonist N-acetylaspartylglutamate (NAAG) into glutamate (Thomas et al. 2000). The resultant increase in NAAG would protect cells from excitotoxic damage. There is an important side effect, however, since the NMDAr inhibitor. NAAG is functionally analogous to the disassociative hallucinogens ketamine, phencyclidine, and nitrous oxide (Jevtovic-Todorovic et al. 2001). These NMDAr antagonists produce a variety of states that may be characterized as either schizophrenomimetic or mystical, such as out-of-body and near-death experiences (Vollenweider et al. 1997). These results suggest that a variety of neurochemicals may mediate some of the more intense experiences associated with meditation practices.

6.9   Autonomic-Cortical Activity

In the early 1970s, Gellhorn and Kiely developed a model of the physiological processes involved in meditation practices based almost exclusively on ANS activity, which while somewhat limited, indicated the importance of the ANS during such experiences (Gellhorn and Kiely 1972). These authors suggested that intense stimulation of either the sympathetic or parasympathetic system, if continued, could ultimately result in simultaneous discharge of both systems (what might be considered a "breakthrough" of the other system). Several studies have demonstrated predominant parasympathetic activity during meditation associated with decreased heart rate and blood pressure, decreased respiratory rate, and decreased oxygen metabolism (Travis 2001). However, a recent study of two separate meditative techniques suggested a mutual activation of parasympathetic and sympathetic systems by demonstrating an increase in the variability of heart rate during meditation (Peng et al. 1999). The increased variation in heart rate was hypothesized to reflect activation of both arms of the ANS. This notion also fits the characteristic description of the MM state in which there is a sense of calmness as well as significant awareness. Also, the notion of mutual activation of both arms of the ANS is consistent with recent developments in the study of autonomic interactions (Hugdahl 1996).

6.10   Serotonergic Activity

Activation of the ANS can result in stimulation of structures in the lateral hypothalamus and median forebrain bundle, which are known to produce both ecstatic and blissful feelings when directly stimulated (Olds and Forbes 1981). Stimulation of the lateral hypothalamus can also result in changes in serotonergic activity. In fact, several studies have shown that after meditation, the breakdown products of serotonin (ST) in urine are significantly increased suggesting an overall elevation in ST during meditation. ST is a neuromodulator that densely supplies the visual centers of the temporal lobe, where it strongly influences the flow of visual associations generated by this area. The cells of the dorsal raphe produce and distribute ST when innervated by the lateral hypothalamus (Aghajanian et al. 1987) and also when activated by the PFC (Juckel et al. 1999). A large number of research studies have implicated the ST system in the pathophysiology of depression, although the exact relationship is not clear. For example, selective ST reuptake inhibitor medications are widely used for the treatment of depression. Since MM in particular has been associated with antidepressant effects (Speca et al. 2000), it is possible that such effects are mediated through the serotonergic system. However, such a hypothesis has yet to be fully evaluated.

Increased ST levels can affect several other neurochemical systems. An increase in ST has a modulatory effect on dopamine suggesting a link between the serotonergic and dopaminergic system that may enhance feelings of euphoria (Vollenweider et al. 1999), frequently described during meditative states. ST, in conjunction with the increased glutamate, has been shown to stimulate the nucleus basalis of Meynert to release acetylcholine, which has important modulatory influences throughout the cortex (Manfridi et al. 1999; Zhelyazkova-Savova et al. 1997). Increased acetylcholine in the frontal lobes has been shown to augment the attentional system and in the parietal lobes to enhance orienting without altering sensory input. Although no studies have evaluated the role of acetylcholine in meditation, it appears that this neurotransmitter may enhance the attentional component as well as the orienting response in the face of progressive deafferentation of sensory input into the parietal lobes during meditation. Increased ST combined with lateral hypothalamic innervation of the pineal gland may lead the latter to increase production of the neurohormone melatonin (MT) from the conversion of ST (Moller 1992). Melatonin has been shown to depress the central nervous system and reduce pain sensitivity (Shaji and Kulkarni 1998).

6.11 Conclusion

There is a growing literature base elucidating neurobiological effects of MM and similar other meditation practices. Much work still needs to be done to synthesize a comprehensive model of the intricate mechanisms underlying MM practices. Furthermore, knowledge of neurotransmitter systems is highly complex and continually being refined. However, the neurophysiological effects that have been observed during meditative states seem to outline a pattern of changes involving certain key cerebral structures in conjunction with autonomic and hormonal changes, paving the way to better understand the effects reported across numerous populations.

The study of meditation, and in particular, MM, may also contribute greatly to the understanding of the larger mind–body problem. Since meditation practices appear to result in substantial effects of both subjective states as well as physiological processes, the study of these practices might be an important area of future research for understanding how the physiological processes in the brain relate to subjective states, mood, cognition, consciousness, and overall human health and well-being. If there are significant reciprocal interactions between the mind and body, then the study of meditation practices may offer a unique and distinct target for future research.

Psychological and clinical effects of meditation are promising and should be taken into consideration by professionals of mental health (Cahn and Polich 2006; Leite et al. 2010). Perception of the world is subject to individual life histories affecting sensibility to specific stimuli, criteria for choice and observation thresholds (Peres et al. 2011). Since meditators foster a state of "willful passivity" in their mind and their relationship to environment, mental health professionals should be aware of possible meditation coping strategies in order to decrease anxiety disorders symptoms or effect compensatory adaptations in response to stressor events. Future improved neurobiological examination methods and researches may contribute to enlighten the phenomenon of different states of consciousness such as those induced by meditation that may impact quality of life.

References

Adair, K. C., Gilmore, R. L., Fennell, E. B., Gold M., & Heilman, K. M. (1995). Anosognosia during intracarotid barbiturate anaesthesia: Unawareness or amnesia for weakness. *Neurology, 45*, 241–243.

Aghajanian, G., Sprouse, J., & Rasmussen, K. (1987). Physiology of the midbrain serotonin system. In H. Meltzer (Ed.), *Psychopharmacology, the third generation of progress* (pp. 141–149). New York: Raven Press.

Albin, R., & Greenamyre, J. (1992). Alternative excitotoxic hypotheses. *Neurology, 42*, 733–738.

Andrews, T. J., Halpern, S. D., & Purves, D. (1997). Correlated size variations in human visual cortex, lateral geniculate nucleus, and optic tract. *Journal of Neuroscience, 17*, 2859–2868.

Armony, J. L., & LeDoux, J. E. (2000). In M. S. Gazzaniga (Ed.), *The new cognitive neurosciences* (pp. 1073–1074). Cambridge: MIT Press.

Baer, R. A. (2003). Mindfulness training as a clinical intervention: A conceptual and empirical review. *Clinical Psychology: Science and Practice, 10*, 125–143.

Baerentsen, K. B., Stødkilde-Jørgensen, H., Sommerlund, B., Hartmann T., Damsgaard-Madsen J., Fosnaes M., & Green, A.C. (2010). An investigation of brain processes supporting meditation. *Cognitive Processing, 11*(1), 57–84.

Brefczynaki-Lewis, J. A., Lutz, A., Schaefer H. S, Levinson D. B., & Davidson, R. J. (2007). Neural correlates of attentional expertise in long-term meditation practitioners. *Proceedings of the National Academy of Sciences of the USA, 104*, 11483–11488.

Bucci, D. J., Conley, M., & Gallagher, M. (1999). Thalamic and basal forebrain cholinergic connections of the rat posterior parietal cortex. *Neuroreport, 10*, 941–945.

Cahn, B. R., & Polich, J. (2006). Meditation states and traits: EEG, ERP, and neuroimaging studies. *Psychological Bulletin, 132*(2), 180–211.

Cheramy, A., Romo, R., & Glowinski, J. (1987). Role of corticostriatal glutamatergic neurons in the presynaptic control of dopamine release. In Sandler, M., Feuerstein, C., & Scatton, B.(Eds.), *Neurotransmitter interactions in the basal ganglia*. New York: Raven.

Chow, T. W., & Cummings, J. L. (1999). In B. L. Miller, & J. L. Cummings (Eds.), *The human frontal lobes* (pp. 3–26). New York: Guilford Press.

Cornwall, J., & Phillipson, O. T. (1988). Mediodorsal and reticular thalamic nuclei receive collateral axons from prefrontal cortex and laterodorsal tegmental nucleus in the rat. *Neuroscience Letters, 88*, 121–126.

Creswell, J. D., Way, B. M., Eisenberger, N. I., & Lieberman, M. D. (2007). Neural correlates of dispositional mindfulness during affect labeling. *Psychosomatic Medicine, 69*, 560–565.

Davies, E., Keyon, C. J., & Fraser, R. (1985). The role of calcium ions in the mechanism of ACTH stimulation of cortisol synthesis. *Steroids, 45*, 557.

Davis, M. (1992). The role of the amygdala in fear and anxiety. *Annual Review of Neuroscience, 15*, 353–375.

Destexhe, A., Contreras, D., & Steriade, M. (1998). Mechanisms underlying the synchronizing action of corticothalamic feedback through inhibition of thalamic relay cells. *Journal of Neurophysiology, 79*, 999–1016.

Dietrich, A. (2003). Functional neuroanatomy of altered states of consciousness: The transient hypofrontality hypothesis. *Consciousness and Cognition, 12*, 231–256.

Elias, A. N., Guich, S., & Wilson, A. F. (2000). Ketosis with enhanced GABAergic tone promotes physiological changes in transcendental meditation. *Medical Hypotheses, 54*, 660–662.

Fernandez-Duque, D., & Posner, M. I. (2001). Brain imaging of attentional networks in normal and pathological states. *Journal of Clinical and Experimental Neuropsychology, 23*, 74–93.

Fish, D. R., Gloor, P., Quesney, F. L., & Olivier, A. (1993). Clinical responses to electrical brain stimulation of the temporal and frontal lobes in patients with epilepsy. *Brain, 116*, 397–414.

Foote, S. (1987). Extrathalamic modulation of cortical function. *Annual Review of Neuroscience, 10*, 67–95.

Frith, C. D., Friston, K., Liddle, P. F., & Frackowiak, R. S. (1991). Willed action and the prefrontal cortex in man a study with PET. *Proceedings of the Royal Society of London, 244*, 241–246.

Gellhorn, E., & Kiely, W. F. (1972). Mystical states of consciousness: Neurophysiological and clinical aspects. *The Journal of Nervous and Mental Disease, 154*, 399–405.

Herzog, H., Lele, V. R., Kuwert, T., Langen K. J, Rota Kops E, & Feinendegen, L. E. (1990–1991). Changed pattern of regional glucose metabolism during Yoga meditative relaxation. *Neuropsychobiol, 23*, 182–187.

Hugdahl, K. (1996). Cognitive influences on human autonomic nervous system function. *Current Opinion in Neurobiology, 6*, 252–258.

Infante, J. R., Peran, F., Martinez, M., Roldan, A., Poyatos, R., Ruiz, C., Samaniego, F., & Garrido, F. (1998). ACTH and beta-endorphin in transcendental meditation. *Physiology and Behavior, 64*, 311–315.

Infante, J. R., Torres-Avisbal, M., Pinel, P., Vallejo, J. A., Peran, F., Gonzalez, F., Contreras, p., Pacheco, C., Roldan, A., & Latre, J.M. (2001). Catecholamine levels in practitioners of the transcendental meditation technique. *Physiology and Behavior, 72*, 141–146.

Ingvar, D. H. (1994). The will of the brain: Cerebral correlates of willful acts. *Journal of Theoretical Biology, 171*, 7–12.

Ivanovski, B., & Malhi, G. S. (2007). The psychological and neurophysiological concomitants of mindfulness forms of meditation. *Acta Neuropsychiatrica, 19*, 76–91.

Janal, M. N., Colt, E. W., Clark, W. C., Glusman, M. (1984). Pain sensitivity, mood and plasma endocrine levels in man following long-distance running: Effects of naxalone. *Pain, 19*, 13–25.

Jevning, R., Wallace, R. K., & Beidebach, M. (1992). The physiology of meditation: A review. A wakeful hypometabolic integrated response. *Neuroscience and Biobehavioral Reviews, 16*, 415–424.

Jevning, R., Wilson, A. F., & Davidson, J. M. (1978). Adrenocortical activity during meditation. *Hormones and Behavior, 10*, 54–60.

Jevtovic-Todorovic, V., Wozniak, D. F., Benshoff, N. D., & Olney, J. W. (2001). A comparative evaluation of the neurotoxic properties of ketamine and nitrous oxide. *Brain Research, 895*, 264–267.

Joseph, R. (1996). *Neuropsychology, neuropsychiatry, and behavioral neurology* (p. 197). New York: Williams & Wilkins.

Juckel, G. J., Mendlin, A., & Jacobs, B. L. (1999). Electrical stimulation of rat medial prefrontal cortex enhances forebrain serotonin output: Implications for electroconvulsive therapy and transcranial magnetic stimulation in depression. *Neuropsychopharmacology, 21*, 391–398.

Kakigi, R., Nakata, H., Inui, K., Hiroe, N., Nagata, O., Honda, M., Tanaka, S., Sadato, N., & Kawakami, M. (2005). Intracerebral pain processing in a Yoga Master who claims not to feel pain during meditation. *European Journal of Pain, 9*(5), 581–589.

Karnath, H. O., Ferber, S., & Himmelbach, M. (2001). Spatial awareness is a function of the temporal not the posterior parietal lobe. *Nature, 411*, 950–953.

Kiss, J., Kocsis, K., Csaki, A., Gorcs, T. J., & Halasz, B. (1997). Metabotropic glutamate receptor in GHRH and beta-endorphin neurons of the hypothalamic arcuate nucleus. *Neuroreport, 8*, 3703–3707.

Kjaer, T. W., Bertelsen, C., Piccini, P., Brooks, D., Alving, J., & Lou, H. C. (2002). Increased dopamine tone during meditation-induced change of consciousness. *Brain Research. Cognitive Brain Research, 13*(2), 255–259.

Lazar, S. W., Bush, G., Gollub, R. L., Fricchione, G. L, Khalsa, G., & Benson, H. (2000). Functional brain mapping of the relaxation response and meditation. *Neuroreport, 11*, 1581–1585.

Leite, J. R., Ornellas, F. L., Amemiya, T. M., de Almeida, A. A., Dias, A. A., Afonso, R., Little, S., & Kozasa, E.H. (2010). Effect of progressive self-focus meditation on attention, anxiety, and depression scores. *Perceptual and Motor Skills, 110*(3 Pt 1), 840–848.

Lieberman, M. D., Eisenberger, N. I., Crocket, M. J., Tom, S. M., Pfeifer, J. H., & Way, B. M. (2007). Putting feelings into words: Affect labeling disrupts amygdale activity in response to affective stimuli. *Psychological Science, 18*, 421–428.

Livesey, J. H., Evans, M. J., Mulligan, R., & Donald, R. A. (2000). Interactions of CRH, AVP and cortisol in the secretion of ACTH from perifused equine anterior pituitary cells: "Permissive" roles for cortisol and CRH. *Endocrine Research, 26*, 445–463.

Lou, H. C., Kjaer, T. W., Friberg, L., Wildschiodtz, G., Holm, S., & Nowak, M. (1999). A 15O-H2O PET study of meditation and the resting state of normal consciousness. *Human Brain Mapping, 7*, 98–105.

Lutz, A., Greischar, L. L., Perlman, D. M., & Davidson, R. J. (2009). BOLD signal in insula is differentially related to cardiac function during compassion meditation in experts vs. novices. *NeuroImage, 47*(3), 1038–1046.

Lynch, J. C. (1980). The functional organization of posterior parietal association cortex. *Behavior Brain Science, 3*, 485–499.

Manfridi, A., Brambilla, D., & Mancia, M. (1999). Stimulation of NMDA and AMPA receptors in the rat nucleus basalis of Meynert affects sleep. *American Journal of Physiology, 277*, R1488–R1492.

Moller, M. (1992). Fine structure of pinealopetal innervation of the mammalian pineal gland. *Microscopy Research and Technique, 21*, 188–204.

Monti, D. A., Peterson, C., Kunkel, E. J., Hauck, W. W., Pequignot, E., Rhodes, L., & Brainard, G. C. (2006). A randomized controlled trial of mindfulness-based art therapy for women with cancer. *Psycho-Oncology, 15*(5), 363–373.

Mountcastle, V. B., Motter, B. C., & Anderson, R. A. (1980). Some further observations on the functional properties of neurons in the parietal lobe of the waking monkey. *Brain Behavior Science, 3*, 520–523.

Newberg, A., Alavi, A., Baime, M., Pourdehnad, M., Santanna, J., & d'Aquili, E. (2001). The measurement of regional cerebral blood flow during the complex cognitive task of meditation: A preliminary SPECT study. *Psychiatry Research: Neuroimaging, 106*, 113–122.

Newberg, A. B., & Iversen, J. (2003). The neural basis of the complex mental task of meditation: Neurotransmitter and neurochemical considerations. *Medical Hypotheses, 61*(2), 282–291.

Newberg, A., Pourdehnad, M., Alavi, A., & d'Aquili, E. (2003). Cerebral blood flow during meditative prayer: Preliminary findings and methodological issues. *Perceptual and Motor Skills, 97*, 625–630.

Newman, J., & Grace, A. A. (1999). Binding across time: The selective gating of frontal and hippocampal systems modulating working memory and attentional states. *Consciousness and Cognition, 8*, 196–212.

O'Halloran, J. P., Jevning, R., Wilson, A. F., Skowsky, R., Walsh, R. N., & Alexander, C. (1985). Hormonal control in a state of decreased activation: Potentiation of arginine vasopressin secretion. *Physiology and Behavior, 35*, 591–595.

Olds, M. E., & Forbes, J. L. (1981). The central basis of motivation, intracranial self-stimulation studies. *Annual Review of Psychology, 32*, 523–574.

Orme-Johnson, D. W., Schneider, R. H., Son, Y. D., Nidich, S., & Cho, Z. H. (2006). Neuroimaging of meditation's effect on brain reactivity to pain. *Neuroreport, 17*(12), 1359–1363.

Palmer, S. E. (1999). *Vision science: Photons to phenomenology*. Cambridge: MIT Press.

Pardo, J. V., Fox, P. T., & Raichle, M. E. (1991). Localization of a human system for sustained attention by positron emission tomography. *Nature, 349*, 61–64.

Peng, C. K., Mietus, J. E., Liu, Y., Khalsa, G., Douglas, P. S., Benson, H., & Goldberger, A. L. (1999). Exaggerates heart rate oscillations during two meditation techniques. *International Journal of Cardiology, 70*, 101–107.

Peres, J. F., Foerster, B., Santana, L. G., Fereira, M. D., Nasello, A. G., Savoia, M., Moreira-Almeida, A., & Lederman, H. (2011). Police officers under attack: Resilience implications of an fMRI study. *Journal of Psychiatric Research, 45*(6), 727–734.

Peres, J., Mercante, J., & Nasello, A. G. (2005). Psychological dynamics affecting traumatic memories: Implications in psychotherapy. *Psychology and Psychotherapy, 78*(4), 431–447.

Peres, J. F., Newberg, A. B., Mercante, J. P., Simão, M., Albuquerque, V. E., Peres, M. J., & Nasello, A. G. (2007). Cerebral blood flow changes during retrieval of traumatic memories before and after psychotherapy: A SPECT study. *Psychological Medicine, 37*(10), 1481–1491.

Peterson, J., Loizzo, J., & Charlson, M. (2009). A program in contemplative self-healing: Stress, allostasis, and learning in the Indo-Tibetan tradition. *Annals of the New York Academy of Sciences, 1172,* 123–147.

Pietrowsky, R., Braun, D., Fehm, H. L., Pauschinger, P., & Born, J. (1991). Vasopressin and oxytocin do not influence early sensory processing but affect mood and activation in man. *Peptides, 12,* 1385–1391.

Poletti, C. E., & Sujatanond, M. (1980). Evidence for a second hippocampal efferent pathway to hypothalamus and basal forebrain comparable to fornix system: A unit study in the monkey. *Journal of Neurophysiology, 44,* 514–531.

Portas, C. M., Rees, G., Howseman, A. M., Josephs, O., Turner, R., & Frith, C. D. (1998). A specific role for the thalamus in mediating the interaction attention and arousal in humans. *Journal of Neuroscience, 18,* 8979–8989.

Ramachandran, V. S., Armel, C., & Foster, C. (1998). Object recognition can drive motion perception. *Nature, 395,* 852–853.

Ramachandran, V. S., & Gregory, R. L. (1991). Perceptual filling in of artificially induced scotomas in human vision. *Nature, 350,* 699–702.

Redding, F. K. (1967). Modification of sensory cortical evoked potentials by hippocampal stimulation. *Electroencephalograph and Clinical Neurophysiology, 22,* 74–83.

Renaud, L. P. (1996). CNS pathways mediating cardiovascular regulation of vasopressin. *Clinical and Experimental Pharmacology and Physiology, 23,* 157–160.

Saver, J. L., & Rabin, J. (1997). The neural substrates of religious experience. *Journal of Neuropsychiatry and Clinical Neurosciences, 9,* 498–510.

Shaji, A. V., & Kulkarni, S. K. (1998). Central nervous system depressant activities of melatonin in rats and mice. *Indian Journal of Experimental Biology, 36,* 257–263.

Sim, M. K., & Tsoi, W. F. (1992). The effects of centrally acting drugs on the EEG correlates of meditation. *Biofeedback and Self-Regulation, 17,* 215–220.

Speca, M., Carlson, L. E., Goodey, E., & Angen, M. (2000). A randomized, wait-list controlled clinical trial: The effect of a mindfulness meditation-based stress reduction program on mood and symptoms of stress in cancer outpatients. *Psychosomatic Medicine, 62*(5), 613–622.

Sudsuang, R., Chentanez, V., & Veluvan, K. (1991). Effects of Buddhist meditation on serum cortisol and total protein levels, blood pressure, pulse rate, lung volume an reaction time. *Physiology and Behavior, 50,* 543–548.

Thomas, A. G., Vornov, J. J., Olkowski, J. L., Merion, A. T., & Slusher, B. S. (2000). N-Acetylated alpha-linked acidic dipeptidase converts N-acetylaspartylglutamate from a neuroprotectant to a neurotoxin. *Journal of Pharmacology and Experimental Therapeutics, 295,* 16–22.

Travis, F. (2001). Autonomic and EEG patterns distinguish transcending from other experiences during transcendental meditation practice. *International Journal of Psychophysiology, 42,* 1–9.

Van Bockstaele, E. J., & Aston-Jones, G. (1995). Integration in the ventral medulla and coordination of sympathetic, pain and arousal functions. *Clinical and Experimental Hypertension, 17,* 153–165.

Vogt, B. A., Finch, D. M., & Olson, C. R. (1992). Functional heterogeneity in cingulate cortex: The anterior executive and posterior evaluative regions. *Cerebral Cortex, 2,* 435–443.

Vollenweider, F. X., Leenders, K. L., Scharfetter, C., Antonini, A., Maguire, P., Missimer, J., & Angst, J. (1997). Metabolic hyperfrontality and psychopathology in the ketamine model of psychosis using positron emission tomography (PET) and [18F]fluorodeoxyglucose (FDG). *European Neuropsychopharmacology, 7,* 9–24.

Vollenweider, F. X., Vontobel, P., Hell, D., & Leenders, K. L. (1999). 5-HT modulation of dopamine release in basal ganglia in psilocybin-induced psychosis in man – a PET study with [11C]raclopride. *Neuropsychopharmacology, 20,* 424–433.

Walton, K. G., Pugh, N. D., Gelderloos, P., & Macrae, P. (1995). Stress reduction and preventing hypertension: Preliminary support for a psychoneuroendocrine mechanism. *Journal of Alternative and Complementary Medicine, 1,* 263–283.

Waterhouse, B. D., Moises, H. C., & Woodward, D. J. (1998). Phasic activation of the locus coeruleus enhances responses of primary sensory cortical neurons to peripheral receptive field stimulation. *Brain Research, 790*, 33–44.

Weingartner, H., Gold, P., Ballenger, J. C., Smallberg, S. A., Summers, R., Rubinow, D. R., Post, R. M., & Goodwin, F. K. (1981). Effects of vasopressin on human memory functions. *Science, 211*, 601–603.

Yadid, G., Zangen, A., Herzberg, U., Nakash, R., & Sagen, J. (2000). Alterations in endogenous brain beta-endorphin release by adrenal medullary transplants in the spinal cord. *Neuropsychopharmacology, 23*, 709–716.

Zhelyazkova-Savova, M., Giovannini, M. G., & Pepeu, G. (1997). Increase of cortical acetylcholine release after systemic administration of chlorophenylpiperazine in the rat: An in vivo microdialysis study. *Neuroscience Letters, 236*, 151–154.

Ziegler, D. R., Cass, W. A., & Herman, J. P. (1999). Excitatory influence of the locus coeruleus in hypothalamic-pituitary-adrenocortical axis responses to stress. *Journal of Neuroendocrinology, 11*, 361–369.

Chapter 7
Functional Neuroimaging Studies of Emotional Self-Regulation and Spiritual Experiences

Mario Beauregard

Abstract In the first section of this chapter, I introduce the concepts of emotional self-regulation and spiritual experiences (SEs). Then, I review the findings of functional neuroimaging studies carried out with regard to the self-regulation of various types of emotional states. In the third section, I examine data suggesting a role for the temporal lobes in SEs, and consider the possibility of experimentally inducing such experiences by stimulating this brain area with weak electromagnetic currents. Next, I review neuroimaging studies of SEs conducted to date. Finally, in the last section, I discuss the theoretical implications of these neuroimaging studies regarding the relationships between subjective experience, mental processes, neurophysiological processes, and human behavior. I also discuss the neuroimaging findings presented and the phenomenology of SEs with respect to the mind-brain problem.

7.1 Introduction

7.1.1 Emotional Self-Regulation

One of the central tenets of evolutionary psychology is that emotions represent efficient modes of adaptation to changing environmental demands (Tooby and Cosmides 1990; Levenson 1994). Emotions, however, are not always the most appropriate response to the various situations encountered in daily life. Indeed, negative emotional states represent one of the main causes of human suffering. This is why it is paramount for us to properly control and modulate our emotional reactions using our cognitive abilities. Fortunately, we humans are capable of emotional

M. Beauregard, PhD (✉)
Department of Psychology, University of Montreal, C.P. 6128, Succursale Centre-Ville,
Montreal, QC, Canada, H3C 3J7
e-mail: mario.beauregard@umontreal.ca

A. Moreira-Almeida and F.S. Santos (eds.), *Exploring Frontiers of the Mind-Brain Relationship*, Mindfulness in Behavioral Health, DOI 10.1007/978-1-4614-0647-1_7,
© Springer Science+Business Media, LLC 2012

self-regulation, i.e., we can change the way we feel by changing the way we think. Without this cognitive capacity, we would be slaves of our emotional impulses and thus unable to interact socially in an adequate and harmonious manner.

Emotional self-regulation refers to the heterogeneous set of cognitive processes by which emotions are self-regulated, i.e., the ways we influence which emotions we have, when we have them, and how we experience and express these emotions (Gross 1999). This capacity is closely associated with self-consciousness, metacognition, and self-agency. This form of self-regulation involves conscious and volitional changes in one or more of the various components (cognitive, experiential, physiological, and behavioral) of emotion (Gross 1999). The cognitive strategies used to self-regulate emotion are numerous and include, among others, reappraisal and cognitive distancing. A growing body of literature indicates that successful regulation of emotion is critical to maintaining physical and mental health (Dennolet et al. 2007).

Several functional neuroimaging studies of emotional self-regulation have been conducted in recent years. These studies support the view that the subjective nature and the intentional content (what they are "about" from a first-person perspective) of the mental events and processes (e.g., volitions, thoughts, feelings) involved in emotional self-regulation significantly influence brain activity.

In the second section of this chapter, I review the findings of functional neuroimaging studies carried out with regard to the self-regulation of sexual arousal, sadness, and negative emotional state. In the last section, I discuss the theoretical implications of these neuroimaging studies regarding the relationships between subjective experience, mental processes, neurophysiological processes, and human behavior.

7.1.2 Spiritual Experiences

The past decade has witnessed an increasing interest in understanding the brain mechanisms mediating spiritual experiences (SEs). These experiences extend or lie beyond the limits of ordinary experience. Mystical experiences represent one particularly interesting kind of SEs. Characterized by altered or expanded consciousness, mystical experiences relate to a fundamental dimension of human existence, and are frequently reported across all cultures and religious/spiritual traditions (Hardy 1975; Hay 1990). For James (1902), the chief characteristics of a mystical experience are as follows: (1) Ineffability: the quality of eluding any adequate account in words; (2) Noetic quality: it is experienced as a state of deep knowledge or insight unknown to the discursive intellect; (3) Transiency: this experience cannot be sustained for long; (4) Passivity: the feeling that, after the experience sets in, one is no longer in control and is perhaps even in the grasp of a superior power. According to Stace (1960), mystical experiences involve the apprehension of an ultimate nonsensuous unity in all things, a oneness or a One to which neither the senses nor the reason can penetrate. Stace further proposes that the main aspects of mystical experiences are as follows: (1) The disappearance of all the mental objects of ordinary consciousness, and the emergence of a unitary or pure consciousness; (2) A sense of objectivity or reality; (3) Feelings of peace, bliss, and joy; (4) The feeling of having encountered the sacred or the divine (sometimes identified as "God"); (5) A trancendence of space and time.

SEs can be triggered by the ingestion of mind-altering substances, shamanic practices, hypnosis (Cardena 2005), and near-death experiences (NDEs). They can also result from regular contemplative practices (prayer, meditation). SEs sometimes occur without any apparent reason (Levin and Steele 2005). These experiences commonly lead to profound transformative changes in attitudes and behavior, i.e., changes in one's worldview, belief system, relationships, and sense of self (Stace 1960; Waldron 1998).

In the third section of this chapter, I examine data suggesting a role for the temporal lobes in SEs, and consider the possibility of experimentally inducing such experiences by stimulating the temporal lobes with weak electromagnetic currents. In the next section, I review neuroimaging studies of SEs conducted to date. Finally, in the last section, I discuss the neuroimaging findings presented and the phenomenology of SEs with regard to the mind–brain problem.

7.2 Neuroimaging Studies of Emotional Self-Regulation

7.2.1 Volitional Control of Sexual Arousal and Sadness

In the first neuroimaging study of conscious and volitional regulation of an emotional state (Beauregard et al. 2001), my group used functional magnetic resonance imaging (fMRI) to identify the neural correlates of the self-regulation of sexual arousal. Ten healthy male volunteers (university students) were scanned during two experimental conditions, i.e., a sexual arousal condition and a down-regulation condition. In the sexual arousal condition, volunteers viewed a series of erotic film excerpts. They were instructed to react normally to these stimuli, i.e., they had to allow themselves to become sexually aroused during the viewing of the erotic film excerpts. In the down-regulation condition, volunteers were instructed to use cognitive distancing, that is, to become a detached observer of comparable erotic film excerpts and the sexual arousal induced by these stimuli. This metacognitive strategy shares some similarity with mindfulness, a central mental state in Buddhist forms of meditation (Thera 2000). To assess the emotional reactions of the volunteers to the film excerpts, they were asked at the outset of each condition to rate on a numerical (analogue) rating scale the intensity of primary emotions felt during the viewing of the film segments.

Phenomenologically, the viewing of the erotic film excerpts during both conditions induced a state of sexual arousal in all volunteers. In the down-regulation condition, most volunteers reported having been successful at distancing themselves from the erotic film excerpts and the sexual arousal induced by these stimuli. Consistent with this, the mean level of sexual arousal was significantly higher in the

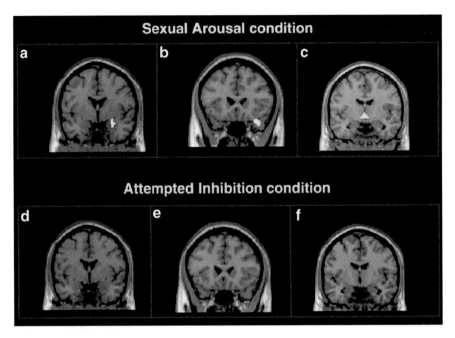

Fig. 7.1 Statistical maps showing activation of limbic–paralimbic structures. In the sexual arousal condition, greater activation during the viewing of erotic film excerpts relative to the viewing of emotionally neutral film excerpts was noted in the right amygdala (**a**), right anterior temporal pole (**b**), and the hypothalamus (**c**). In the suppression condition, no significant loci of activation were seen in the amygdalae (**d**), the anterior temporal polar region (**e**), and the hypothalamus (**f**) (reproduced with permission of the Society for Neuroscience from Beauregard et al. 2001)

sexual arousal condition than in the down-regulation condition. Additionally, in line with the results of a previous study (Karama et al. 2002), the viewing of the erotic film excerpts in the sexual arousal condition produced a significant activation of the right amygdala, right anterior temporal pole (Brodmann area [BA] 38), and hypothalamus (Fig. 7.1). In the down-regulation condition, activation peaks were noted in BA 10 of the right lateral prefrontal cortex (LPFC) and BA 32 of the right rostroventral anterior cingulate cortex (ACC). Of note, no significant locus of blood oxygenation level-dependent (BOLD) signal increase was measured in the amygdala, anterior temporal pole, and hypothalamus (Fig. 7.2).

These results provide robust evidence for the view previously proposed that emotional self-regulation depends on a brain system in which prefrontal cortical areas mediate the cognitive modulation of emotional responses (Nauta 1971; Tucker et al. 1995; Davidson et al. 2000). Moreover, these results accord with studies indicating that the LPFC is involved in metacognitive/executive top–down processes (these processes refer to the ability to monitor and control the information processing necessary to produce voluntary action – Flavell 1979). This prefrontal cortical region has been implicated in the selection and control of behavioral strategies and action (Fuster 1999), especially in the inhibition of inherent response tendency

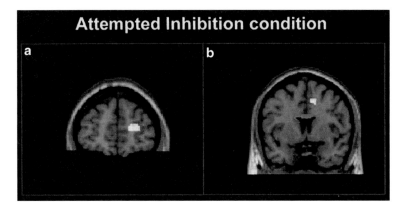

Fig. 7.2 Statistical maps showing peaks of activation in various prefrontal cortical regions during the *Attempted Inhibition condition*. Images are coronal sections for the data averaged across subjects. The right hemisphere of the brain corresponds to the *right side* of the image. Significant loci of activation were noted in the right superior frontal gyrus (**a**) and right anterior cingulate gyrus (**b**) (reproduced with permission of the Society for Neuroscience from Beauregard et al. 2001)

(Goldman-Rakic 1987; Damasio 1995; Frith and Dolan 1996; Fuster 1997). These findings also support the hypothesis that by virtue of its anatomic connections with brain regions implicated in the modulation of autonomic and endocrine functions (e.g., amygdala, hypothalamus, and orbitofrontal cortex [OFC]), the rostro-ventral subdivision of the ACC plays a key role in the regulation of the autonomic aspect of emotional responses (Vogt et al. 1992; Devinsky et al. 1995; Bush et al. 2000). The fact that this portion of the ACC has both afferent and efferent connections with neural structures mediating autonomic functions such as the periacqueductal gray, the dorsal motor nucleus of the vagus, and the preganglionic sympathetic neurons in the intermediolateral cell column of the spinal cord (Benes 1997) supports this hypothesis.

A protocol similar to that used in the Beauregard et al. (2001) study was utilized in another fMRI study carried out by our research group (Lévesque et al. 2003). This study sought to delineate the neural substrates of conscious and voluntary regulation of sadness, a primary emotion with a negative valence (Plutchik 1994). Twenty healthy female volunteers were scanned during a sad condition, in which participants viewed sad film excerpts, and a down-regulation condition, in which participants had to become detached observers of similar film excerpts. Experientially, the mean level of reported sadness was significantly higher in the sad condition than in the down-regulation condition. Neurally, significant bilateral loci of activation were measured during the sad condition in the anterior temporal pole (BA 21 and 38) and midbrain. Significant BOLD signal increases were also seen in the right ventro-lateral prefrontal cortex (VLPFC) (BA 47), left amygdala, and left insula. In the down-regulation condition, significant loci of BOLD activations were detected in the right LPFC (BA 9) and right OFC (BA 11).

The right orbitofrontal activation noted during down-regulation of sadness is consistent with the fact that the OFC (BA 11) is located at the junction of the prefrontal associative cortex and the so-called limbic system. Since this cortical area sends projections to the amygdala, anterior temporal pole, insular cortex, LPFC, and brain stem, it has been associated with the integration of visceral-autonomic processes with cognitive and behavioral processes (Ongür et al. 1998; Rempel-Clower and Barbas 1998; Eslinger 1999a). It has been further proposed that this portion of the OFC plays a central role in the regulation of socio-emotional behavior in settings involving self-consciousness, inhibition, and the self-guidance of behavior through judgments and decisions about one's actions (Dolan 1999; Eslinger 1999b; Fuster 1997; Grafman and Litvan 1999).

Altogether, the results of our fMRI studies indicate that conscious and voluntary regulation of various types of emotional responses is mediated by a few areas of the PFC, such as the LPFC, the OFC, and the rostro-ventral ACC.

7.2.2   Conscious and Volitional Regulation of Negative Emotional State

To date a number of fMRI studies have been carried out regarding the neural substrates of self-regulation of negative emotional state. In one of these studies, Schaefer et al. (2002) used an event-related design to test the hypothesis that volitional down-regulation of emotionally negative pictures is associated with changes in neural activity within the amygdala, a cardinal component of emotion perception. Negative and emotionally neutral pictures selected from the International Affective Picture System (IAPS) (Lang et al. 1998) were presented to seven healthy female volunteers. Volunteers were instructed to either maintain the initial emotional response induced by the picture throughout its presentation or passively view the picture without self-regulating the emotional reaction. After each picture presentation, volunteers indicated how they currently felt using a Likert-type scale. Volunteers reported feeling more negative during negatively picture trials than neutral picture trials. Volunteers also reported feeling more negative on maintain trials than view trials. In keeping with previous functional neuroimaging studies having shown increased activation in the amygdala in response to emotionally negative stimuli (Phan et al. 2002), greater amygdalar signal change was noted during the presentation of negative pictures compared with neutral pictures. Increased activation in the amygdala was associated with maintenance of negative emotion compared with the passive viewing condition. Furthermore, a prolonged BOLD signal increase was found in the amygdala when volunteers maintained the negative emotional state during the presentation of negative pictures. This amygdalar signal increase was significantly correlated with volunteers' self-reported levels of negative emotion. These findings suggest that the conscious cognitive processes that modulated the emotional responses of the volunteers to the negative pictures were associated with an alteration of the degree of neural activity within the amygdala. In other words, the degree of amygdalar activity can be consciously and volitionally regulated.

In another event-related fMRI study, Ochsner et al. (2002) used reappraisal as a cognitive strategy to modulate emotional experience (reappraisal consists of altering the trajectory of an unfolding emotional response by cognitively reinterpreting/transforming the meaning of the emotion-eliciting stimulus/event to change one's emotional response to it). Negative and neutral pictures from the IAPS were presented to 15 healthy female volunteers. For each trial volunteers were instructed to view the picture and allow themselves to experience/feel any emotional response it might elicit. The picture remained on the screen for an additional period of time with an instruction either to attend or reappraise. On attend trials, volunteers were requested to attend to and be aware of, but not to try to modify, any feelings induced by negative or neutral pictures. On reappraise trials, volunteers were instructed to reinterpret the negative pictures so that they no longer generated a negative emotional response. After the presentation of each picture volunteers had to rate on a four-point scale the strength of current negative emotional state. Behaviorally, reappraisal of negative pictures successfully lessened negative emotion. The average ratings of the strength of negative emotion were significantly lower on reappraise trials than on attend trials. Neurofunctionally, reappraisal was associated with a significant activation of the dorsal and ventral regions of the left LPFC (BA 6, 8, 10, 44, 46) as well as the dorsal medial prefrontal cortex (MPFC) (BA 8). Moreover, the right amygdala was significantly more activated on attend than reappraise trials. Interestingly, the medial OFC (MOFC) (BA 11) displayed greater activation to most negative pictures on attend than on reappraise trials, whereas activated areas of the LPFC showed the opposite trend. For one of these areas, the VLPFC, increased activation during reappraisal was correlated across volunteers with decreased activation in amygdala.

For Ochsner et al. (2002), these results provide good evidence that reappraisal may modulate the emotion processes implemented in the amygdala and MOFC and involved in the evaluation of the emotional significance and contextual relevance of a stimulus or event. The specific LPFC and MPFC areas found associated with reappraisal were comparable to the prefrontal areas frequently seen activated in diverse working memory and response-selection tasks that entail resisting to interference from competing stimuli and maintaining information in awareness (Courtney et al. 1998; Smith and Jonides 1999; Cabeza and Nyberg 2000). This similitude supports the view that a common ensemble of prefrontal areas underlie the cognitive regulation of both feelings and thoughts (Knight et al. 1999; Smith and Jonides 1999; Miller and Cohen 2001; Ochsner and Feldman Barrett 2001).

Ochsner et al. (2004) conducted another fMRI study in 24 healthy female volunteers to compare the neural systems supporting down and up-regulation with reappraisal of negative emotion induced by IAPS pictures. Results indicated that amygdala activation was modulated up or down in agreement with the regulatory goal. Up-regulation recruited the ACC (BA 32), MPFC (BA 9, 32), left LPFC (BA 9), and left amygdala whereas down-regulation recruited the left LPFC (BA 8, 9) and the left lateral OFC (BA 47). For Ochsner et al. (2004), these results suggest that both common and distinct neural systems underlie various forms of reappraisal.

More recently, Phan et al. (2005) scanned 14 healthy male and female volunteers while they down-regulated – via reappraisal – negative IAPS pictures. Online subjective ratings of intensity of negative emotion were used as covariates of brain activity. Down-regulation of negative emotion was associated with activation of dorsal ACC (BA 32), right dorsal MPFC (BA 9), and right LPFC (BA 9), as well as attenuation of brain activity within the left amygdala vicinity. Furthermore, activity within right dorsal ACC (BA 32) was negatively correlated with intensity of negative emotion, whereas activation of the amygdala was positively correlated with intensity of negative emotion. These findings highlight a functional dissociation of corticolimbic brain responses implicating enhanced activation of PFC and attenuation of limbic areas during down-regulation of negative emotion.

Other fMRI studies (Kim and Hamann 2007; Goldin et al. 2008) have confirmed the implication of the LPFC (BA 9, 10), ACC (BA 24, 32), MPFC (BA 9, 10), and OFC (BA 11) in the reappraisal of negative pictures or film clips. These data corroborate the view that several prefrontal cortical areas are involved in the cognitive regulation of emotion.

7.3 Role of the Temporal Lobe and the Limbic System in Spiritual Experiences

7.3.1 Temporal Lobe Epilepsy

Clinical observations suggest an association between temporal-lobe epilepsy (TLE) and SEs during (ictal), after (postictal), and in between (interictal) seizures (Devinsky and Lai 2008). Howden (1872–1873) first observed a man who had a religious conversion after a generalized seizure in which he experienced being transported to "Heaven." Following this report, Mabille (1899) described a patient who, following a seizure, mentioned that God had given him a mission to bring law to the world. A few years later, Spratling (1904) reported a religious aura or a premonitory period of hours or several days associated with religiosity in 52 of 1,325 patients with epilepsy (4%). Additionally, Boven (1919) described the case of a 14-year-old boy who, after a seizure, recounted having seen God and angels.

More recently, Dewhurst and Beard (1970) reported six patients with TLE who underwent sudden and often lasting religious conversions in the postictal period. Some of these patients had prior or active psychiatric disorders. There was an obvious temporal relationship between conversion and first seizure or increased seizure frequency in five patients.

Studies have shown that between 0.4 and 3.1% of TLE patients had ictal SEs while postictal SEs occurred in 2.2% of patients with TLE. Ictal SEs occur most often in patients with right TLE whereas there is a predominance of postictal and interictal SEs in TLE patients with bilateral seizure foci. It is noteworthy that many of the epilepsy-related religious conversion experiences occur postictally (Devinsky and Lai 2008).

From an experiential perspective, ictal religious experiences during seizures can be accompanied by intense emotions of God's presence, the sense of being connected to the infinite (Alajouanine 1963), hallucinations of God's voice (Hansen and Brodtkorb 2003), visual hallucination of a religious figure (Karagulla and Robertson 1955), or repetition of a religious phrase (Ozkara et al. 2004). It has been suggested that some of the greatest religious figures in history (e.g., Saint Paul, Muhammad, Joan of Arc, Joseph Smith) were probably suffering from TLE (Saver and Rabin 1997).

Naito and Matsui (1988) described an elderly woman whose seizures were characterized by joyful visions of God. Interictal electroencephalography (EEG) revealed spike discharges in the left anterior and middle temporal areas during sleep. Moreover, Morgan (1990) reported the case of a patient whose seizures were associated with feelings of ineffable contentment and fulfillment, visualizing a bright light recognized as the source of knowledge, and sometimes visualizing a bearded young man resembling Jesus Christ. A computed axial tomography (CAT) scan displayed a right anterior temporal astrocytoma. Following anterior temporal lobectomy, ecstatic seizures vanished. Along the same lines, Picard and Craig (2009) described the case of a 64-year-old right-handed woman who has had epileptic seizures with an ecstatic aura. During her ecstatic epileptic seizures, she reported experiencing immense joy above physical sensations as well as unimaginable harmony with life, the world and the "All." Cerebral magnetic resonance imaging (MRI) showed a meningioma in the left temporal pole region. An interictal EEG revealed left anterior temporofrontal epileptiform activity.

Ogata and Miyakawa (1998) examined 234 Japanese epileptic patients for ictus-related religious experiences. Three (1.3%) patients were found to have had such experiences. All three cases had TLE with post-ictal psychosis, and interictal experiences with hyperreligiosity. Patients who had ictus-related or interictal religious experiences did not believe only in Buddhism (a traditional religion in Japan), but also rather in a combination of Buddhism and Shintoism, new Christian sects, contemporary Japanese religions, and/or other folk beliefs. Interestingly the content of their religious experiences were related to their religious beliefs. This finding emphasizes the importance of considering psychological factors – such as beliefs – in addition to neurobiological aspects when the relationship between epilepsy and religion/spirituality is discussed.

Kelly and Grosso (2007) have argued that, generally, symptoms of TLE have not much in common with the various aspects of SEs. Indeed, people having temporal lobe seizures typically report auditory and visual hallucinations, dysphoria, fear, anger, or impressions of depersonalization and unreality. Importantly, in national surveys in the United States, Britain, and Australia, 20–49% of the individuals interviewed reported SEs (Hay 1994). Given the relatively common incidence of SEs in the general population, it is fair to say that most people who have these experiences are not epileptics. If TLE really produced such experiences, all or most people suffering from this disorder would have them.

7.3.2   Interictal Personality Syndrome of Temporal Lobe Epilepsy

Waxman and Geschwind (1975) suggested that hyperreligiosity is a core feature of a distinctive interictal personality syndrome of TLE (also called the Geschwind syndrome). A heightened state of religious conviction, an increased sense of personal destiny, intense philosophical and cosmological concerns, and strong moral beliefs usually characterize interictal religiosity. The putative temporal-lobe personality type is also characterized by hypermoralism, deepened affects, humorlessness, aggressive irritability, and hypergraphia.

Support for this hypothetical syndrome was provided by Bear and Fedio (1977), who found that religiosity trait scores were significantly higher in TLE patients than in healthy control participants. In keeping with this, Roberts and Guberman (1989) found that 60% of 57 consecutive patients with epilepsy had excessive interests in religion. Subsequent studies using religion questionnaires, however, failed to find any differences about interictal religiosity between patients with TLE vs. idiopathic generalized epilepsy, or between patients with epilepsy and normal control participants (Willmore et al. 1980; Tucker et al. 1987). It has been proposed that differences in religiosity measures and in control group selection account for some of the discrepancy among studies (Saver and Rabin 1997).

In other respects, Wuerfel et al. (2004) used MRI to investigate mesial temporal structures in 33 patients with refractory partial epilepsy, comparing 22 patients without and 11 patients with hyper-religiosity. High ratings on the religiosity scale were correlated with a significantly smaller hippocampus in the right hemisphere. The hippocampal atrophy may reflect the duration and severity of hyperreligiosity. This does not mean, however, that the right hippocampus is the critical cerebral structure for religious or spiritual experiences (Devinsky and Lai 2008).

7.3.3   The Limbic-Marker Hypothesis

Saver and Rabin (1997) have theorized that temporolimbic discharges underlie each of the core features of SEs (e.g., the noetic and the ineffable; the sense of having touched the ultimate ground of reality; the sense of incommunicability of the experience; the experience of unity, timelessness, and spacelessness; feelings of positive affect, peace, and joy). The limbic system integrates external stimuli with internal drives and is part of a distributed neural circuit that marks the valence (positive or negative) of stimuli and experiences (Damasio et al. 1991). Saver and Rabin (1997) posited that temporolimbic discharges may mark experiences as (1) depersonalized or derealized, (2) crucially important and self-referent, (3) harmonious-indicative of a connection or unity between disparate elements, and (4) ecstatic-profoundly joyous.

According to the limbic-marker hypothesis, the perceptual and cognitive contents of a SE are comparable to those of ordinary experience, except that they are tagged by the limbic system as of deep importance, as united into a whole, and/or as joyous. Therefore, descriptions of the contents of the SE resemble descriptions of the contents of ordinary experience, and the feelings associated with them cannot be captured fully in words. As in the case of strong emotions, these limbic markers can be named but cannot be communicated in their full visceral intensity, resulting in a report of ineffability.

The temporal lobe and the limbic system may not be the only cerebral structures involved in SEs. About this question, Devinsky and Lai (2008) hypothesized that alterations in frontal functions in the right hemisphere may contribute to increased religious interests and beliefs as a personality trait. This hypothesis is based on the finding that dramatic changes in self, defined as a change in political, social, or religious views can be seen in patients with a dysfunction affecting selectively the right frontal lobe (Miller et al. 2001).

7.3.4   Stimulation of the Temporal Lobe

Persinger (1983) speculated that SEs are evoked by transient, electrical microseizures within deep structures of the temporal lobes, and that it is possible to experimentally induce SEs by stimulating the temporal lobes with weak electromagnetic currents. Persinger and Healey (2002) tested this hypothesis by exposing 48 undergraduate psychology students to weak (100 nT to 1 µT) complex, pulsed electromagnetic fields. These fields were applied in one of three ways: over the right temporoparietal region, over the left temporoparietal region, or equally across the temporoparietal region of both hemispheres of the brain (one treatment per group). Fields were applied for 20 min while participants were wearing opaque goggles in a very quiet room. A fourth group was exposed to a sham field condition – that is, participants were not exposed to an electromagnetic field, although all participants were told that they might be. Beforehand, the Hypnosis Induction Profile (Spiegel et al. 1976) was administered to participants, to test for suggestibility.

Two-thirds of the participants reported a "sensed presence" – the sense that someone else was with them – under the influence of the electromagnetic fields. But 33% of the control (sham-field) group reported a sensed presence too. That is, Persinger and Healey (2002) found that twice as many participants reported a sensed presence under the influence of the electromagnetic field as those who reported one without an electromagnetic field. About half of these participants described a sentient being who moved when they tried to "focus attention" upon the presence, and approximately one-third of participants attributed the presence to a deceased member of the family or to some cultural equivalent of a "spirit guide." In the study, those who had received stimulation over the right hemisphere or both hemispheres reported more unusual phenomena than those who had received stimulation over the left hemisphere.

Persinger and Healey (2002) concluded two things: that the experience of a sensed presence can be manipulated by experiment, and that such an experience may be the source for phenomena attributed to visitations by spiritual entities.

A research team at Uppsala University in Sweden, headed by Granqvist et al. (2005), mirrored Persinger's experiment by testing 89 undergraduate students, some of whom were exposed to the electromagnetic fields and some of whom were not. Using Persinger's equipment, the Swedish researchers could not reproduce his key results. They attributed their findings to the fact that they ensured that neither the participants nor the experimenters interacting with them had any idea who was being exposed to the electromagnetic fields.

Granqvist and colleagues made sure that their experiment was double blind by using two experimenters for each trial. The first experimenter, who was not told about the purpose of the study, interacted with the participants. The second experimenter switched electromagnetic fields off or on without advising either the first experimenter or the subject. So if the subject had not already been advised that a SE was likely at Granqvist's laboratory, the study experimenters were not in a position to provide that clue.

Study participants included undergraduate theology students as well as psychology students. Neither group was asked for prior information on spiritual or paranormal experiences, nor was any participant told that there was a sham-field (control) condition. Rather, participants were told only that the study investigated the influence of weak electromagnetic fields on experiences and feeling states. Personality characteristics that might predispose a person to report an unusual experience were used as predictors for which participants would report one. These characteristics included absorption (the ability to become completely absorbed in an experience), signs of abnormal temporal-lobe activity, and a "New Age" lifestyle orientation.

No evidence was found for a "sensed presence" effect of weak electromagnetic fields. The characteristic that significantly predicted the outcomes was personality. Of the three participants who reported strong SEs, two were members of the control group. Of the 22 who reported "subtle" experiences, 11 were members of the control group. Those participants who were rated as highly suggestible on the basis of a questionnaire filled out after they completed the study reported paranormal experiences whether the electromagnetic field was on or off while they were wearing the stimulation helmet. Granqvist and colleagues also noted that they had found difficult to evaluate the reliability of Persinger's findings, because no information on experimental randomization or blindness was provided, which left his results open to the possibility that psychological suggestion was the best explanation.

The brief descriptions of the experiences reported by the participants in Persinger's studies bear very little resemblance to genuine SEs. These descriptions do not support his claim that he and his colleagues have been able to induce authentic SEs by applying weak electromagnetic fields over the temporal lobes. Such conclusion should not come as a surprise given that studies involving direct electrical stimulation of the temporal lobes in epileptic patients have failed to produce SEs (Kelly and Grosso 2007). As a matter of fact, direct electrical stimulation of the temporal lobes elicits very rarely any mental response, such as sensations, images, thoughts, and emotional feelings (Halgren et al. 1978). This does not mean, however, that the temporal lobe is not implicated in the neural mediation of SEs.

7.4 Neuroimaging Studies of Spiritual Experiences

The first brain imaging study of a religious experience was conducted by Azari et al. (2001). These researchers studied a group of six self-identified religious participants, who attributed their religious experience to biblical Psalm 23. These participants, who were members of a "Free Evangelical Fundamentalist Community" in Germany, all reported having had a conversion experience (related to the first verse of biblical Psalm 23, which states "The LORD is my shepherd; I shall not be in want"), and interpreted biblical text literally as the word of God. Religious participants were compared to six nonreligious individuals. The texts used for the different tasks were "religious" (first verse of biblical Psalm 23), "happy" (a well-known German children's nursery rhyme), and "neutral" (instructions on using a phone card from the Düsseldorf telephone book).

Participants were scanned with positron emission tomography (PET) during various conditions: reading silently or reciting biblical Psalm 23; reading silently or reciting the children's nursery rhyme; reading silently the set of instructions; and while lying quietly. The PET images revealed a significant activation of the right dorsolateral prefrontal cortex in the religious participants during the religious condition (relative to other readings) when compared with nonreligious participants. During the religious state, the religious participants showed additional loci of activation, including the dorsomedial frontal cortex and the right precuneus. Limbic areas did not show regional cerebral blood flow (rCBF) changes.

According to Azari et al. (2001), these results support the view that religious experience is a cognitive attributional phenomenon, mediated by a pre-established neural circuit, involving dorsolateral prefrontal, dorsomedial frontal, and medial parietal cortex. Religious attributions are based on religious schemata which consist in organized knowledge about religion and religious issues, and include reinforced structures for inferring religiously related causality of experienced events (Spilka and McIntosh 1995). Azari et al. (2001) proposed that the dorsolateral prefrontal and medial parietal cortices were probably involved in the subject's own religious schemata whereas the dorsomedial frontal cortex was implicated in the felt immediacy of religious experience.

Newberg et al. (2003) used single photon emission computed tomography (SPECT) to scan three Franciscan nuns while they performed a "centering prayer" to open themselves to the presence of God. This prayer involved the internal repetition of a particular phrase. Compared with baseline, the prayer condition scan showed increased rCBF in the prefrontal cortex, inferior parietal lobes, and inferior frontal lobes. There was a strong inverse correlation between the rCBF changes in the prefrontal cortex and in the ipsilateral superior parietal lobule. Newberg et al. (2003) hypothesized that increased frontal rCBF reflected focused concentration whereas changed rCBF in the superior parietal lobule was related to an altered sense of space experienced by the nuns during prayer. In this pilot study, there was no attempt to analyze and quantify in a rigorous and systematic manner the nuns' subjective experiences during their "centering prayer." That is, Newberg and colleagues could not determine whether focusing attention on a phrase from a prayer over a period of time really led the nuns to feel the presence of God.

Newberg et al. (2006) also used SPECT to investigate changes in cerebral activity during glossolalia ("speaking in tongues"). This unusual mental state is associated with specific religious traditions. Glossolalia is one of the "gifts of the Spirit" according to Saint Paul and, hence, some fundamentalist religious traditions see it as a sign of being visited by the Spirit. This is due to the Pentecost experience, where, according to the Acts of the Apostles, the Apostles "spoke in the tongues" of all those present, i.e., made themselves understood to everybody, whereby later on just babbling something became synonymous with glossolalia. In this state, the individual seems to be speaking in an incomprehensible language over which he/she claims to have no voluntary control. Yet, the individual perceives glossolalia to have great personal and religious meaning. In their study, Newberg and colleagues examined five practitioners (women) of glossolalia. Participants described themselves as Christians in a Charismatic or Pentecostal tradition who had practiced glossolalia for more than 5 years. Structured clinical interviews excluded current psychiatric conditions. Glossolalia was compared with a religious singing state since the latter is similar except that it involves actual language (English). Earphones were used to play music to sing and to perform glossolalia (the same music was used for both conditions). Several significant rCBF differences were noted between the glossolalia and singing states. During glossolalia (compared with the religious singing state), significant decreases were found in the prefrontal cortices, left caudate, and left temporal pole. Decreased activity in the prefrontal lobe is consistent with the participants' description of a lack of volitional control over the performance of glossolalia. Newberg et al. (2006) proposed that the decrease in the left caudate may relate to the altered emotional activity during glossolalia.

Recently we sought to identify the neural correlates of a mystical experience (as understood in the Christian sense) in a group of contemplative nuns using fMRI (Beauregard and Paquette 2006). Fifteen Carmelite nuns took part in the study. BOLD signal changes were measured during a mystical condition, a control condition, and a baseline condition. In the mystical condition, participants were asked to remember and relive the most intense mystical experience ever felt in their lives as a member of the Carmelite Order. This strategy was adopted given that the nuns told us before the onset of the study that "God can't be summoned at will." In the control condition, participants were instructed to remember and relive the most intense state of union with another human ever felt in their lives while being affiliated with the Carmelite Order. The week preceding the experiment, participants were requested to practice these two tasks. The baseline condition was a normal restful state.

Immediately at the end of the scan, the intensity of the subjective experience during the control and mystical conditions was measured using numerical rating scales ranging from 0 (no experience of union) to 5 (most intense experience of union ever felt): self-report data referred solely to the experiences lived during these two conditions, not to the original experiences recalled to self-induce the control and mystical states. The phenomenology of the mystical experience during the mystical condition was assessed with the Mysticism Scale (Hood 1975).

In regard to the phenomenology of the subjective experience during the mystical condition, several participants mentioned that they felt the presence of God, His unconditional and infinite love, as well as plenitude and peace. All participants reported that from a first-person perspective, the experiences lived during the mystical condition were different than those used to self-induce a mystical state. Participants also reported the presence of visual and motor imagery during both the mystical and control conditions. Additionally, the participants experienced a feeling of unconditional love during the control condition.

The mystical vs. baseline contrast produced significant loci of BOLD activation in the right MOFC, right middle temporal cortex, right inferior parietal lobule and superior parietal lobule, right caudate, left MPFC, left dorsal ACC, left inferior parietal lobule, left insula, left caudate, and left brainstem. A few loci of activation were also seen in the extra-striate visual cortex.

Based on the studies indicating a relationship between SEs and the temporal lobe, we posited that the right middle temporal activation noted during the mystical condition was related with the subjective impression of contacting a spiritual reality. We also proposed that the caudate activations reflected feelings of joy and unconditional love since the caudate nucleus has been systematically activated in previous functional brain imaging studies implicating positive emotions such as happiness (Damasio et al. 2000), romantic love (Bartels and Zeki 2000), and maternal love (Bartels and Zeki 2004). Concerning the brainstem, there is some empirical support for the view that certain brainstem nuclei map the organism's internal state during emotion (Damasio 1999). Given this, it is conceivable that the activation in the left brainstem was linked to the somatovisceral changes associated with the feelings of joy and unconditional love. As for the insula, this cerebral structure is richly interconnected with regions involved in autonomic regulation (Cechetto 1994). It contains a topographical representation of inputs from visceral, olfactory, gustatory, visual, auditory, and somatosensory areas and is proposed to integrate representations of external sensory experience and internal somatic state (Augustine 1996). The insula has been seen activated in several studies of emotional processing and appears to support a representation of somatic and visceral responses accessible to consciousness (Damasio 1999; Critchley et al. 2004). It is plausible that the left insular activation noted in our study was related to the representation of the somatovisceral reactions associated with the feelings of joy and unconditional love.

In addition, we suggested that the left medial prefrontal cortical activation was linked with conscious awareness of those feelings. Indeed, the results of functional neuroimaging studies indicate that the MPFC is involved in the metacognitive representation of one's own emotional state (Lane and Nadel 2000). This prefrontal area receives sensory information from the body and the external environment via the OFC and is heavily interconnected with limbic structures, such as the amygdala, ventral striatum, hypothalamus, midbrain periaqueductal gray region, and brainstem nuclei (Barbas 1993; Carmichael and Price 1995). In other respects, brain imaging findings (Lane et al. 1997, 1998) support the view that the activation of the left dorsal ACC reflected that aspect of emotional awareness associated with the interoceptive detection of emotional signals during the mystical condition. This cortical region projects strongly to the visceral regulation areas in the hypothalamus and midbrain periaqueductal gray (Ongür et al. 2003). Regarding the MOFC, there is mounting evidence that this prefrontal cortical region codes for subjective pleasantness (Kringelbach et al. 2003). The MOFC has been found activated with regard to the pleasantness of the taste or smell of stimuli (de Araujo et al. 2003; Rolls et al. 2003) or music (Blood and Zatorre 2001). It has reciprocal connections with the cingulate and insular cortices (Carmichael and Price 1995; Cavada et al. 2000). The right medial orbitofrontal cortical activation noted in the mystical condition was perhaps related to the fact that the experiences lived during the mystical state were considered by the participants emotionally pleasant.

Given that the right superior parietal lobule is involved in the spatial perception of self (Neggers et al. 2006), it is conceivable that the activation of this parietal area reflected a modification of the body schema associated with the impression that something greater than the participants seemed to absorb them. Moreover, there is evidence that the left inferior parietal lobule is part of a neural system implicated in the processing of visuospatial representation of bodies (Felician et al. 2003). Therefore, the left inferior parietal lobule activation in the mystical condition was perhaps related to an alteration of the body schema. In keeping with this, there is some evidence indicating that the right inferior parietal lobule is crucial in bodily consciousness and the process of self/other distinction (Ruby and Decety 2003). However, the inferior parietal lobule plays an important role in motor imagery (Decety 1996). It is thus possible that the activations in the right and left inferior parietal lobules were related to the motor imagery experienced during the mystical condition. Lastly, about the loci of activation found in the extra-striate visual cortex during this condition, it has been previously shown (Ganis et al. 2004) that this region of the brain is implicated in visual mental imagery. It is likely that the BOLD activation in visual cortical areas was related to the visual mental imagery reported by the nuns.

We also used EEG to measure spectral power and coherence in the Carmelite nuns during the same type of mystical state (Beauregard and Paquette 2008). EEG activity was recorded from 19 scalp locations during a resting state, a control condition, and a mystical condition. In the mystical condition compared with the control condition, electrode sites showed greater theta power at F3, C3, P3, Fz, Cz, and Pz, and greater gamma1 power was detected at T4 and P4. Higher delta/beta ratio, theta/alpha ratio, and theta/beta ratio were found for several electrode sites. Additionally, FP1-C3 pair of electrodes displayed greater coherence for theta band while F4-P4, F4-T6, F8-T6, and C4-P4 pairs of electrodes showed greater coherence for alpha band. These results indicate that mystical experiences are mediated by marked changes in EEG power and coherence. These changes implicate several cortical areas in both hemispheres (Beauregard and Paquette 2008).

Taken together, the results of our fMRI and EEG studies of the Carmelite nuns dispose of the notion that there is a "God spot" in the temporal lobes of the brain that can somehow "explain" SEs (Ramachandran and Blakeslee 1998). These results further suggest that SEs are complex and multidimensional experiences, which are mediated by different brain regions involved in a variety of functions, such as self-consciousness, emotion, body representation, mental imagery, and the perception of contacting another reality.

7.5 Implications with Regard to the Mind–Brain Problem

7.5.1 Emotional Self-Regulation

The neuroimaging studies of emotional self-regulation reviewed in this chapter indicate that the conscious and voluntary use of cognitive distancing and reappraisal selectively modulates the way the brain responds to emotional stimuli (Beauregard 2007). These neuroimaging investigations strongly support the mentalist view that the subjective nature and the intentional content of mental processes and states significantly influence the functioning and plasticity of the brain, that is, mental variables can be causally efficacious and explanatory and have to be considered as much as neurophysiological variables to reach a correct understanding of human behavior. Such a mentalist outlook is supported by the high explanatory and predictive value of agentic factors such as volitions, goals, beliefs, and expectations (Bandura 2001).

Findings from neuroimaging studies of emotional self-regulation call into question the psychophysical identity theory and epiphenomenalism. For the psychophysical identity theory, mental processes are identical with neural processes (Feigl 1958) whereas for epiphenomenalism, mental processes are causally inert epiphenomena (i.e., by-products) of neural processes. These findings also challenge eliminative materialism (or eliminativism). According to this view, mental processes, which are pre-scientific concepts belonging to unsophisticated ideas of how the brain works, can be reduced entirely to brain processes. Eliminative materialism further proposes that all common language or "folk psychology" descriptions of mental experience should be eliminated and replaced by descriptions using neuroscientific language (Churchland 1981). For these materialist views, physically describable brain mechanisms represent the core and final explanatory vehicle for every kind of psychologically described data. These views are extremely counterintuitive since our most basic experience teaches us that our choice of perspective about how we apprehend our mental processes and states makes a huge difference in how we respond to them (Schwartz et al. 2005).

I stand firmly against the inclination of certain neuroscientists and philosophers toward neuroreductionism, i.e., the reduction of human beings to their brains (a form of "neural anthropomorphism"), and posit that the brain is necessary but not sufficient to explicate all the human psychological features. This position is shared by Fuchs (2008) and Glannon (2009). In my view, persons are conscious, perceive, think, feel emotion, interpret, believe and make decisions, not parts of their brains. To attribute such capacities to brains is to commit a philosophical mistake that Bennett and Hacker (2003) identified as "the mereological fallacy" in neuroscience, i.e., the fallacy of attributing to parts of the brain attributes that are properties of the whole human person.

With the emergence of self-consciousness, self-agency, and metacognitive capacities, evolution has enabled us to consciously and volitionally shape the functioning of our brains. These advanced capacities allow us to be driven not only by survival and reproduction but also by complex sets of insights, goals, and beliefs. For example, the ethical values associated with a given spiritual tradition help certain people to keep in check their emotional impulses and behave in an altruistic fashion. In this particular instance, moral conscience replaces innate programming as behavioral regulator, and permits emancipation from "selfish" genes and the primitive dictates of the mammalian brain.

7.5.2 Spiritual Experiences

Physicalism is the mainstream metaphysical view of modern neuroscience with respect to the mind–body problem, i.e., the explanation of the relationship that exists between mental processes and bodily processes. According to this view, consciousness and mental events (e.g., thoughts, emotions, desires) can be reduced to their neural correlates, i.e., the brain electrical and chemical processes whose presence necessarily and regularly accompanies these mental events. Physicalist philosophers and neuroscientists believe that mental events are equivalent to brain processes. Standing against this metaphysical belief, James (1898) notes that neural correlates of SEs do not yield a causal explanation of mental events, i.e., they cannot explain how neural processes become mental events. Indeed correlation does not entail causation. And the external reality of "God" or ultimate reality can neither be confirmed nor disconfirmed by neural correlates.

Newberg et al. (2001) submitted that the most important criterion for judging what is real is the subjective vivid sense of reality. They argued that individuals usually refer to dreams as less real than waking (baseline) reality when they are recalled within baseline reality. In contrast, SEs (e.g., "cosmic consciousness" states, religious visions, transcendent aspects of NDEs) appear more real to the experiencers than waking (baseline) reality when they are recalled from baseline reality.

A major problem with this criterion is its subjectivity. This problem is well illustrated by the fact that individuals suffering from psychosis are unable to distinguish personal subjective experience from the reality of the external world. They experience hallucinations as being very real. From a neuroscientific point of view, a more satisfactory approach to evaluate the "objective" ontological reality of SEs is to determine whether it is possible for a human being to have such an experience during a state of clinical death, i.e., when her/his brain is not functioning. In this state, vital signs have ceased: the heart is in ventricular fibrillation, there is a total lack of electrical activity on the cortex of the brain (flat EEG), and brainstem activity is abolished (loss of the corneal reflex, fixed and dilated pupils, and loss of the gag reflex).

The thought-provoking case of a patient who apparently underwent a profound SE while her brain was not functioning has been reported by cardiologist Michael Sabom (1998). In 1991, 35-year-old Atlanta-based singer and songwriter Pam Reynolds began to suffer dizziness, loss of speech, and difficulty moving. A CAT scan revealed that she had a giant basilar artery aneurysm (a grossly swollen blood vessel in the brainstem). If it burst, it would kill her. But attempting to drain and repair it might kill her too. Her physician offered no chance of survival using conventional procedures. Reynolds heard about neurosurgeon Robert Spetzler, at the Barrow Neurological Institute in Phoenix, Arizona. He is a specialist and pioneer in a rare and dangerous technique called hypothermic cardiac arrest, or "Operation Standstill." Spetzler would take her body down to a temperature so low that she was clinically dead, but then bring her back to a normal temperature before irreversible damage set in. At a low temperature, the swollen vessels that burst at the high temperatures needed to sustain human life become soft. Then they can be operated upon with less risk. Furthermore, the cooled brain can survive longer without oxygen, though it obviously cannot function in that state. So for all practical purposes, Reynolds would actually be clinically dead during the surgery. But if she did not agree to it, she would soon be dead anyway with no hope of return. So she consented.

Reynolds was brought into the operating room at 7:15 a.m. She was given general anesthesia to produce a rapid loss of awareness. Her eyes were lubricated to prevent drying and then taped shut. EEG electrodes were used to monitor the electrical activity of her cerebral cortex. The activity of her brainstem was measured repeatedly through 100-dB clicks emitted by small molded speakers inserted into her ears. Assisted by over 20 physicians, nurses, and technicians, Spetzler began at 8:40 a.m. the surgery by cutting through Reynolds's skull with a surgical saw. At that point, she reported that she felt herself "pop" outside her body and hover above. From her out-of-body position, she could see the doctors working on her body. She observed, "I thought the way they had my head shaved was very peculiar. I expected them to take all of the hair, but they did not" (Sabom 1998, p. 41). She could also perceive the Midas Rex bone saw used to open skulls, and the sound made by the saw ("I hated the sound of looked like an electric tootbrush....") (Sabom 1998, p. 41). She later described this saw with considerable accuracy (Sabom 1998).

The outermost membrane of Reynolds's brain was then removed by Spetzler and cut open with scissors. At about the same time, a female cardiac surgeon attempted to locate the femoral artery in Reynolds's right groin. The blood vessels being too small to accept the abundant blood flow necessitated by the cardiopulmonary bypass machine, the left femoral artery was prepared instead by this cardiac surgeon.

Interestingly, Reynolds later claimed to remember a female voice saying "We have a problem. Her arteries are too small." And then a male voice: "Try the other side."

At 10:50 a.m., a tube was inserted into the left femoral artery. This tube was connected to the cardiopulmonary bypass machine. Then Reynolds's body temperature began to fall as the warm blood circulated from the artery into the cylinders of the bypass machine, where it was cooled down before being returned to her body. Reynolds's heart stopped 15 min later. Her EEG brain waves flattened into total silence (during a cardiac arrest, the brain's electrical activity disappears after 10–20 s – Clute and Levy 1990). A few minutes later, her brainstem became totally unresponsive, and her body temperature fell to a sepulchral 60°F. At 11:25 a.m., the head of the operating table was titled up, the bypass machine was turned off, and the blood was drained from her body. Pam Reynolds was clinically dead.

During this period, she became conscious of floating out of the operating room and traveling down a tunnel with a light. Deceased relatives and friends were waiting at the end of this tunnel, including her long-dead grandmother. She entered the presence of a brilliant, wonderfully warm and loving light and sensed that her soul was part of God and that everything in existence was created from the light (the breathing of God). This extraordinary experience ended when Reynolds's deceased uncle led her back to her body. She compared reentering her body to "plunging into a pool of ice."

The aneurysm was removed by Spetzler when all the brain had drained from Reynolds's brain. Then, the bypass machine was turned on, and warm blood was pumped into her body. As her body temperature started to increase, the brainstem began to respond to the clicking speakers in her ears, and the EEG recorded electrical activity in the cortex. The bypass machine was turned off at 12:32 p.m. Reynolds was taken to the recovery room in stable condition at 14:10.

Pam Reynolds described her NDE as a continuous experience perceived to be as real at the beginning, during her out-of-body experience (OBE), as it was throughout. She estimated that the NDE ended at the close of the surgery.

Skeptics will argue that when Reynolds saw the surgeon cutting her skull or heard a female voice say something about the size of her blood vessels, she was not clinically dead yet. Nevertheless, her ears were blocked by small molded speakers continuously emitting 100-dB clicks (100 dB correspond approximately to the noise produced by a speeding express train). Medical records confirmed that these words were effectively pronounced (Sabom 1998). Moreover, these speakers were affixed with tape and gauze. It is thus highly unlikely that Reynolds could have physically overheard operating room conversation.

Reynolds's case does not seem to be the only one of its type. Indeed, we recently found another case, also of a young woman who underwent a hypothermic cardiac standstill a few years ago. This surgical procedure is used in Montreal by cardiothoracic surgeons to repair aortic arch defects (in the top part of the aorta), which may be lethal. As in the case of Pam Reynolds, this woman reported a deep SE while she was apparently dead clinically (unpublished findings).

Such cases strongly challenge the physicalist doctrine in regard to the mind–brain problem. They suggest that mental processes and events can be experienced at the moment that the brain seemingly no longer functions (as evidenced by a flat EEG and an unresponsive brain stem) during a period of clinical death. These cases also suggest that SEs can occur when the brain is not functioning, that is, these experiences are not necessarily delusions created by a defective brain. In other words, it would be possible for humans to experience a spiritual reality during an altered state of consciousness in which perception, cognition, identity, and emotion function independently from the brain. This raises the possibility that when a SE happens while the brain is fully functional, the neural correlates of this experience indicate that the brain is de facto connecting with a spiritual level of reality. Solid scientific research is required to tackle this fascinating issue.

It should be noted that since Pam Reynolds and the Montreal woman did not die, there were likely residual brain processes not detectable by EEG that persisted during the clinical death period at sufficient levels so as to permit return to normal brain functioning after the standstill operation. Yet it is difficult to see how the brain could generate higher mental functions in the absence of cortical and brainstem activity. Scientific research is clearly needed to investigate the possibility that a functioning brain may not be essential to higher mental functions and SEs. It is noteworthy that NDEs are reported by approximately 10–20% of cardiac arrest survivors (Parnia et al. 2001; van Lommel et al. 2001; Greyson 2003).

More than a century ago, William James (1898) speculated that the brain may serve a permissive/transmissive/expressive function rather than a productive one, in terms of the mental events and experiences it allows (just as a prism – which is not the source of the light – changes incoming white light to form the colored spectrum). In line with James's view, I recently proposed the Psychoneural Translation Hypothesis (or PTH) (Beauregard 2007). This hypothesis posits that the mind (the psychological world, the first-person perspective) and the brain (which is part of the "physical" world, the third-person perspective) represent two epistemologically and ontologically distinct domains that can interact because they are complementary aspects of the same underlying reality. Furthermore, the PTH postulates that conscious and unconscious mental processes and events are selectively translated, based on a specific code, into neural processes and events at the various levels of brain organization (e.g., biophysical, molecular, chemical, neural circuits). Stated otherwise, mentalese (the language of the mind) is translated into neuronese (the language of the brain). This informational transduction mechanism, which has emerged throughout the evolution of the human species, allows mental processes to causally influence brain activity in a very precise manner. The mentalese-to-neuronese translation might be based on quantum processes, as suggested by physicist Henry Stapp. In this view, causal connections between the psychological world and the physical world are specified by the von Neumann interpretation of quantum physics (Schwartz et al. 2005).

Following William James, Henri Bergson (1914) and Aldous Huxley (1954) speculated that the brain acts as a filter (or reducing valve) by blocking out much of, and allowing registration and expression of only a narrow band of, perceivable reality. Bergson and Huxley believed that over the course of evolution, the brain has been trained to eliminate most of those perceptions that do not directly aid our everyday survival. This outlook implies that the brain normally limits our capacity to have a SE. In keeping with this, it is likely that alterations in electrical and chemical activities in the brain are necessary for SEs to take place (Beauregard and O'Leary 2007). Such alterations can be occasioned by a variety of means. Given this, it is probable that distinct patterns of brain activity can mediate SEs.

References

Alajouanine, T. (1963). Dostoiewski's epilepsy. *Brain, 86,* 209–218.

Augustine, J. R. (1996). Circuitry and functional aspects of the insular lobe in primates including humans. *Brain Research Reviews, 22,* 229–244.

Azari, N. P., Nickel, J., Wunderlich, G., Niedeggen, M., Hefter, H., Tellmann, L., et al. (2001). Neural correlates of religious experience. *The European Journal of Neuroscience, 13,* 1649–1652.

Bandura, A. (2001). Social cognitive theory: An agentic perspective. *Annual Review of Psychology, 52,* 1–26.

Barbas, H. (1993). Organization of cortical afferent input to the orbitofrontal area in the rhesus monkey. *Neuroscience, 56,* 841–864.

Bartels, A., & Zeki, S. (2000). The neural basis of romantic love. *Neuroreport, 11,* 3829–3834.

Bartels, A., & Zeki, S. (2004). The neural correlates of maternal and romantic love. *NeuroImage, 21,* 1155–1166.

Bear, D., & Fedio, P. (1977). Quantitative analysis of interictal behavior in temporal lobe epilepsy. *Archives of Neurology, 34,* 454–467.

Beauregard, M. (2007). Mind does really matter: Evidence from neuroimaging studies of emotional self-regulation, psychotherapy, and placebo effect. *Progress in Neurobiology, 81,* 218–236.

Beauregard, M., Lévesque, J., & Bourgouin, P. (2001). Neural correlates of the conscious self-regulation of emotion. *The Journal of Neuroscience, 21,* RC165.

Beauregard, M., & O'Leary, D. (2007). *The spiritual brain.* New York: Harper Collins.

Beauregard, M., & Paquette, V. (2006). Neural correlates of a mystical experience in Carmelite nuns. *Neuroscience Letters, 405,* 186–190.

Beauregard, M., & Paquette, V. (2008). EEG activity in Carmelite nuns during a mystical experience. *Neuroscience Letters, 444,* 1–4.

Benes, F. (1997). Corticolimbic circuitry and the development of psychopathology during childhood and adolescence. In N. A. Krasnegor, G. Reid Lyon, & P. S. Goldman-Rakic (Eds.), *Development of the prefrontal cortex* (pp. 211–240). Baltimore: Paul H. Brookes.

Bennett, M. R., & Hacker, P. M. S. (2003). *Philosophical foundations of neuroscience.* New York: Blackwell.

Bergson, H. (1914). Presidential address. *Proceedings of the Society for Psychical Research, 27,* 157–175.

Blood, A., & Zatorre, R. (2001). Intensely pleasurable responses to music correlate with activity in brain regions implicated in reward and emotion. *Proceedings of the National Academy of Sciences of the United States of America, 98,* 11818–11823.

Boven, W. (1919). Religiosité et épilepsie. *Schweizer Archiv für Neurologie und Psychiatrie, 4,* 153–169.

Bush, G., Luu, P., & Posner, M. I. (2000). Cognitive and emotional influences in anterior cingulate cortex. *Trends in Cognitive Sciences, 4*, 215–222.

Cabeza, R., & Nyberg, L. (2000). Neural bases of learning and memory: Functional neuroimaging evidence. *Current Opinion in Neurology, 13*, 415–421.

Cardena, E. (2005). The phenomenology of deep hypnosis: Quiescent and physically active. *The International Journal of Clinical and Experimental Hypnosis, 53*, 37–59.

Carmichael, S. T., & Price, J. L. (1995). Limbic connections of the orbital and medial prefrontal cortex in macaque monkeys. *The Journal of Comparative Neurology, 363*, 615–641.

Cavada, C., Company, T., Tejedor, J., Cruz-Rizzolo, R. J., & Reinoso-Suarez, F. (2000). The anatomical connections of the macaque monkey orbitofrontal cortex. A review. *Cerebral Cortex, 10*, 220–242.

Cechetto, D. F. (1994). Identification of a cortical site for stress-induced cardiovascular dysfunction. *Integrative Physiological and Behavioral Science, 29*, 362–373.

Churchland, P. M. (1981). Eliminative materialism and the propositional attitudes. *The Journal of Philosophy, 78*, 67–90.

Clute, H. L., & Levy, W. J. (1990). Electroencephalographic changes during brief cardiac arrest in humans. *Anesthesiology, 73*, 821–825.

Courtney, S. M., Petit, L., Maisog, J. M., Ungerleider, L. G., & Haxby, J. V. (1998). An area specialized for spatial working memory in human frontal cortex. *Science, 279*, 1347–1351.

Critchley, H. D., Wien, S., Rotshtein, P., Ohman, A., & Dolan, R. J. (2004). Neural systems supporting interoceptive awareness. *Nature Neuroscience, 7*, 189–195.

Damasio, A. R. (1995). On some functions of the human prefrontal cortex. *Annals of the New York Academy of Sciences, 769*, 241–251.

Damasio, A. R. (1999). *The feeling of what happens: Body and emotion in the making of consciousness*. New York: Harcourt Brace.

Damasio, A. R., Grabowski, T. J., Bechara, A., Damasio, H., Ponto, L. L., Parvizi, J., et al. (2000). Subcortical and cortical brain activity during the feeling of self-generated emotions. *Nature Neuroscience, 3*, 1049–1056.

Damasio, A. R., Tranel, D., & Damasio, H. (1991). Somatic markers and the guidance of behavior. In H. Levin, H. Eisenberg, & A. Benton (Eds.), *Frontal lobe function and dysfunction* (pp. 217–228). New York: Oxford University Press.

Davidson, R. J., Putnam, K. M., & Larson, C. L. (2000). Dysfunction in the neural circuitry of emotion regulation – A possible prelude to violence. *Science, 289*, 591–594.

de Araujo, I. E., Rolls, E. T., Kringelbach, M. L., McGlone, F., & Phillips, N. (2003). Taste-olfactory convergence, and the representation of the pleasantness of flavour, in the human brain. *The European Journal of Neuroscience, 18*, 2059–2068.

Decety, J. (1996). Do imagined and executed actions share the same neural substrate? *Cognitive Brain Research, 3*, 87–93.

Dennolet, J., Nyklicek, I., & Vingerhoets, A. (2007). *Emotion, emotion regulation, and health*. New York: Springer.

Devinsky, O., & Lai, G. (2008). Spirituality and religion in epilepsy. *Epilepsy and Behavior, 12*, 636–643.

Devinsky, O., Morrell, M. J., & Vogt, B. A. (1995). Contributions of anterior cingulate cortex to behavior. *Brain, 118*, 279–306.

Dewhurst, K., & Beard, A. W. (1970). Sudden religious conversions in temporal lobe epilepsy. *The British Journal of Psychiatry, 117*, 497–507.

Dolan, R. J. (1999). On the neurology of morals. *Nature Neuroscience, 11*, 927–929.

Eslinger, P. J. (1999a). Orbital frontal cortex: Historical and contemporary views about its behavioral and physiological significance (part I). *Neurocase, 5*, 225–229.

Eslinger, P. J. (1999b). Orbital frontal cortex: Behavioral and physiological significance (part II). *Neurocase, 5*, 299–300.

Feigl, H. (1958). The "mental" and the "physical". In H. Feigl, M. Scriven, & G. Maxwell (Eds.), *Concepts, theories and the mind-body problem* (pp. 320–492). Minneapolis: Minnesota Studies in the Philosophy of Science.

Felician, O., Ceccaldi, M., Didic, M., Thinus-Blanc, C., & Poncet, M. (2003). Pointing to body parts: A double dissociation study. *Neuropsychologia, 41*, 1307–1316.

Flavell, J. H. (1979). Metacognition and cognitive monitoring: A new area of cognitive development inquiry. *American Psychologist, 34*, 906–911.

Frith, C., & Dolan, R. (1996). The prefrontal cortex in higher cognitive functions. *Cognitive Brain Research, 5*, 175–181.

Fuchs, T. (2008). *Das Gehirn – Ein Beziehungsorgan*. Berlin: Springer.

Fuster, J. M. (1997). *The prefrontal cortex: Anatomy, physiology and neuropsychology of the frontal lobe* (3rd ed.). Philadelphia: Lippincott-Raven.

Fuster, J. M. (1999). Synopsis of function and dysfunction of the frontal lobe. *Acta Psychiatrica Scandinavica, 99*, 51–57.

Ganis, G., Thompson, W. L., & Kosslyn, S. M. (2004). Brain areas underlying visual mental imagery and visual perception: An fMRI study. *Cognitive Brain Research, 20*, 226–241.

Glannon, W. (2009). Our brains are not us. *Bioethics, 23*, 321–329.

Goldin, P. R., McRae, K., Ramel, W., & Gross, J. J. (2008). The neural bases of emotion regulation: Reappraisal and suppression of negative emotion. *Biological Psychiatry, 15*, 577–586.

Goldman-Rakic, P. S. (1987). Circuitry of primate prefrontal cortex and regulation of behavior by representational memory. In V. B. Mountcastle, F. Plum, & S. R. Geiger (Eds.), *Handbook of physiology* (Vol. 5, pp. 373–417). Bethesda: American Physiological Society.

Grafman, J., & Litvan, I. (1999). Importance of deficits in executive functions. *The Lancet, 354*, 1921–1923.

Granqvist, P., Fredrikson, M., Unge, P., Hagenfeldt, A., Valind, S., Larhammar, D., et al. (2005). Sensed presence and mystical experiences are predicted by suggestibility, not by the application of transcranial weak complex magnetic fields. *Neuroscience Letters, 379*, 1–6.

Greyson, B. (2003). Incidence and correlates of near-death experiences in a cardiac care unit. *General Hospital Psychiatry, 25*, 269–276.

Gross, J. J. (1999). Emotion regulation: Past, present, future. *Cognition and Emotion, 13*, 551–573.

Halgren, E., Walter, R. D., Cherlow, D. G., & Crandall, P. H. (1978). Mental phenomena evoked by electrical stimulation of the human hippocampal formation and amygdala. *Brain, 101*, 83–117.

Hansen, B., & Brodtkorb, E. (2003). Partial epilepsy with "ecstatic" seizures. *Epilepsy and Behavior, 4*, 667–673.

Hardy, A. (1975). *The biology of God*. New York: Taplinger.

Hay, D. (1990). *Religious experience today: Studying the facts*. London: Mowbray.

Hay, D. (1994). The biology of God: What is the current status of Hardy's hypothesis? *The International Journal for the Psychology of Religion, 4*, 1–23.

Hood, R. W., Jr. (1975). The construction and preliminary validation of a measure of reported mystical experience. *The Journal for the Scientific Study of Religion, 14*, 21–41.

Howden, J. C. (1872–1873). The religious sentiments in epileptics. *The Journal of Mental Science, 18*, 491–497.

Huxley, A. (1954). *The doors of perception*. New York: Haper and Row.

James, W. (1898). Human immortality: Two supposed objections to the doctrine. In G. Murphy & R. O. Ballou (Eds.), *William James on psychical research* (pp. 279–308). New York: Viking.

James, W. (1902). *The varieties of religious experience: A study in human nature*. New York: Longmans, Green and Co.

Karagulla, S., & Robertson, E. E. (1955). Psychical phenomena in temporal lobe epilepsy and the psychoses. *British Medical Journal, 1*, 748–752.

Karama, S., Lecours, A. R., Leroux, J.-M., et al. (2002). Areas of brain activation in males and females during viewing of erotic film excerpts. *Human Brain Mapping, 16*, 1–13.

Kelly, E. F., & Grosso, M. (2007). Mystical experience. In E. F. Kelly & E. W. Kelly (Eds.), *Irreducible mind: Toward a psychology for the 21st century*. Lanham: Rowman & Littlefield.

Kim, S. H., & Hamann, S. J. (2007). Neural correlates of positive and negative emotion regulation. *Journal of Cognitive Neuroscience, 19*, 776–798.

Knight, R. T., Staines, W. R., Swick, D., & Chao, L. L. (1999). Prefrontal cortex regulates inhibition and excitation in distributed neural networks. *Acta Psychologica, 101*, 159–178.

Kringelbach, M. L., O'Doherty, J., Rolls, E. T., & Andrews, C. (2003). Activation of the human orbitofrontal cortex to a liquid food stimulus is correlated with its subjective pleasantness. *Cerebral Cortex, 13*, 1064–1071.

Lane, R. D., Fink, G. R., Chau, P. M., & Dolan, R. J. (1997). Neural activation during selective attention to subjective emotional responses. *Neuroreport, 8*, 3969–3972.

Lane, R. D., & Nadel, L. (Eds.). (2000). *Cognitive neuroscience of emotion*. New York: Oxford University Press.

Lane, R. D., Reiman, E. M., Axelrod, B., Yun, L. S., Holmes, A., & Schwartz, G. E. (1998). Neural correlates of levels of emotional awareness. Evidence of an interaction between emotion and attention in the anterior cingulate cortex. *Journal of Cognitive Neuroscience, 10*, 525–535.

Lang, P. J., Bradley, M. M., & Cuthbert, B. N. (1998). Emotion and motivation: Measuring affective perception. *Journal of Clinical Neurophysiology, 15*, 397–408.

Levenson, C. A. (1994). Human emotion: A functional view. In P. Ekman & R. J. Davidson (Eds.), *Fundamental questions about the nature of emotion* (pp. 123–126). New York: Oxford University Press.

Lévesque, J., Eugène, F., Joanette, Y., et al. (2003). Neural circuitry underlying voluntary self-regulation of sadness. *Biological Psychiatry, 53*, 502–510.

Levin, J., & Steele, L. (2005). The spiritual experience: Conceptual, theoretical, and epidemiological perspectives. *Explore, 1*, 89–101.

Mabille, H. (1899). Hallucinations religieuses dans l'épilepsie. *Annales Médicopsychologiques, 9–10*, 76–81.

Miller, E. K., & Cohen, J. D. (2001). An integrative theory of prefrontal cortex function. *Annual Review of Neuroscience, 24*, 167–202.

Miller, B. L., Seeley, W. W., Mychack, P., Rosen, H. J., Mena, I., & Boone, K. (2001). Neuroanatomy of the self: Evidence from patients with frontotemporal dementia. *Neurology, 57*, 817–821.

Morgan, H. (1990). Dostoevsky's epilepsy: A case report and comparison. *Surgical Neurology, 33*, 413–416.

Naito, H., & Matsui, N. (1988). Temporal lobe epilepsy with ictal ecstatic state and interictal behavior of hypergraphia. *The Journal of Nervous and Mental Disease, 176*, 123–124.

Nauta, W. J. H. (1971). The problem of the frontal lobe: A reinterpretation. *Journal of Psychiatry Research, 8*, 167–187.

Neggers, S. F., Van der Lubbe, R. H., Ramsey, N. F., & Postma, A. (2006). Interactions between ego- and allocentric neuronal representations of space. *NeuroImage, 31*, 320–331.

Newberg, A., d'Aquili, E., & Rause, V. (2001). *Why God won't go away*. New York: Ballantine Books.

Newberg, A., Pourdehnad, M., Alavi, A., & d'Aquili, E. G. (2003). Cerebral blood flow during meditative prayer: Preliminary findings and methodological issues. *Perceptual and Motor Skills, 97*, 625–630.

Newberg, A. B., Wintering, N. A., Morgan, D., & Waldman, M. R. (2006). The measurement of regional cerebral blood flow during glossolalia: A preliminary SPECT study. *Psychiatry Research, 148*, 67–71.

Ochsner, K. N., Bunge, S. A., Gross, J. J., & Gabrieli, J. D. (2002). Rethinking feelings: An FMRI study of the cognitive regulation of emotion. *Journal of Cognitive Neuroscience, 14*, 1215–1229.

Ochsner, K. N., & Feldman Barrett, L. (2001). A multiprocess perspective on the neuroscience of emotion. In T. J. Mayne & G. Bonanno (Eds.), *Emotions: Current issues and future directions* (pp. 38–81). New York: Guilford.

Ochsner, K. N., Ray, R. D., Cooper, J. C., et al. (2004). For better or for worse: Neural systems supporting the cognitive down- and up-regulation of negative emotion. *NeuroImage, 23*, 483–499.

Ogata, A., & Miyakawa, T. (1998). Religious experiences in epileptic patients with a focus on ictus-related episodes. *Psychiatry and Clinical Neurosciences, 52*, 321–325.

Ongür, D., An, X., & Price, J. L. (1998). Prefrontal cortical projections to the hypothalamus in macaque monkeys. *The Journal of Comparative Neurology, 401*, 480–505.

Ongür, D., Ferry, A. T., & Price, J. L. (2003). Architectonic subdivision of the human orbital and medial prefrontal cortex. *The Journal of Comparative Neurology, 460*, 425–449.

Ozkara, C., Sary, H., Hanoglu, L., Yeni, N., Aydogdu, I., & Ozyurt, E. (2004). Ictal kissing and religious speech in a patient with right temporal lobe epilepsy. *Epileptic Disorders, 6*, 241–245.

Parnia, S., Waller, D. G., Yeates, R., & Fenwick, P. (2001). A qualitative and quantitative study of the incidence, features and aetiology of near death experiences in cardiac arrest survivors. *Resuscitation, 48*, 149–156.

Persinger, M. A. (1983). Religious and mystical experiences as artefacts of temporal lobe function: A general hypothesis. *Perceptual and Motor Skills, 57*, 1255–1262.

Persinger, M. A., & Healey, F. (2002). Experimental facilitation of the sensed presence: Possible intercalation between the hemispheres induced by complex magnetic fields. *The Journal of Nervous and Mental Disease, 190*, 533–541.

Phan, K. L., Fitzgerald, D. A., Nathan, P. J., Moore, G. J., Uhde, T. W., & Tancer, M. E. (2005). Neural substrates for voluntary suppression of negative affect: A functional magnetic resonance imaging study. *Biological Psychiatry, 57*, 210–219.

Phan, K. L., Wager, T., Taylor, S. F., & Liberzon, I. (2002). Functional neuroanatomy of emotion: A meta-analysis of emotion activation studies in PET and fMRI. *NeuroImage, 16*, 331–348.

Picard, F., & Craig, A. D. (2009). Ecstatic epileptic seizures: A potential window on the neural basis for human self-awareness. *Epilepsy and Behavior, 16*, 539–546.

Plutchik, R. (1994). *The psychology and biology of emotion*. New York: Harper Collins College.

Ramachandran, V. S., & Blakeslee, S. (1998). *Phantoms in the brain: Probing the mysteries of the human mind*. New York: Morrow.

Rempel-Clower, N. L., & Barbas, H. (1998). Topographic organization of connections between the hypothalamus and prefrontal cortex in the rhesus monkey. *The Journal of Comparative Neurology, 398*, 393–419.

Roberts, J. K., & Guberman, A. (1989). Religion and epilepsy. *Psychiatry Journal of University of Ottawa, 14*, 282–286.

Rolls, E. T., Kringelbach, M. L., & de Araujo, I. E. (2003). Different representations of pleasant and unpleasant odours in the human brain. *The European Journal of Neuroscience, 18*, 695–703.

Ruby, P., & Decety, J. (2003). What you believe versus what you think they believe: A neuroimaging study of conceptual perspective-taking. *The European Journal of Neuroscience, 17*, 2475–2480.

Sabom, M. (1998). *Light and death: One doctor's fascinating account of near-death experiences*. Grand Rapids: Zondervan.

Saver, J. L., & Rabin, J. (1997). The neural substrates of religious experience. *The Journal of Neuropsychiatry and Clinical Neurosciences, 9*, 498–510.

Schaefer, S. M., Jackson, D. C., Davidson, R. J., Aguirre, G. K., Kimberg, D. Y., & Thompson-Schill, S. L. (2002). Modulation of amygdalar activity by the conscious regulation of negative emotion. *Journal of Cognitive Neuroscience, 14*, 913–921.

Schwartz, J. M., Stapp, H., & Beauregard, M. (2005). Quantum theory in neuroscience and psychology: A neurophysical model of mind/brain interaction. *Philosophical Transactions of the Royal Society of London. Series B, Biological Sciences, 360*, 1309–1327.

Smith, E. E., & Jonides, J. (1999). Storage and executive processes in the frontal lobes. *Science, 283*, 1657–1661.

Spiegel, H., Aronson, M., Fleiss, J. L., & Haber, J. (1976). Psychometric analysis of the Hypnotic Induction Profile. *The International Journal of Clinical and Experimental Hypnosis, 24*, 300–315.

Spilka, B., & McIntosh, D. N. (1995). Attribution theory and religious experience. In R. W. Hood (Ed.), *Handbook of religious experience* (pp. 421–445). Birmingham: Religious Education Press.

Spratling, W. P. (1904). *Epilepsy and its treatment*. Philadelphia: WB Saunders.

Stace, W. T. (1960). *Mysticism and philosophy*. New York: Macmillan.

Thera, N. (2000). *The vision of Dhamma: Buddhist writings of Nyanaponika Thera*. Seattle: BPS Pariyatti Editions.

Tooby, J., & Cosmides, L. (1990). On the universality of human nature and the uniqueness of the individual: The role of genetics and adaptation. *Journal of Personality, 58*, 17–67.

Tucker, D. M., Luu, P., & Pribram, K. H. (1995). Social and emotional self-regulation. *Annals of the New York Academy of Sciences, 769*, 213–239.

Tucker, D. M., Novelly, R. A., & Walker, P. J. (1987). Hyperreligiosity in temporal lobe epilepsy: Redefining the relationship. *The Journal of Nervous and Mental Disease, 175*, 181–184.

van Lommel, P., van Wees, R., Meyers, V., & Elfferich, I. (2001). Near-death experience in survivors of cardiac arrest: A prospective study in the Netherlands. *The Lancet, 358*, 2039–2045.

Vogt, B. A., Finch, D. M., & Olson, C. R. (1992). Functional heterogeneity in cingulate cortex: The anterior executive and posterior evaluative regions. *Cerebral Cortex, 2*, 435–443.

Waldron, J. L. (1998). The life impact of spiritual experiences with a pronounced quality of noesis. *Journal of Transpersonal Psychology, 30*, 103–134.

Waxman, S. G., & Geschwind, N. (1975). The interictal behavior syndrome of temporal lobe epilepsy. *Archives of General Psychiatry, 32*, 1580–1586.

Willmore, L. J., Heilman, K. M., Fennell, E., & Pinnas, R. M. (1980). Effect of chronic seizures on religiosity. *Transactions of the American Neurological Association, 105*, 85–87.

Wuerfel, J., Krishnamoorthy, E. S., Brown, R. J., Lemieux, L., Koepp, M., Tebartz van Elst, L., et al. (2004). Religiosity is associated with hippocampal but not amygdala volumes in patients with refractory epilepsy. *Journal of Neurology, Neurosurgery, and Psychiatry, 75*, 640–642.

Part IV
Human Experiences as Promising lines of Investigation of Mind-Brain Relationship

Chapter 8
Can Near Death Experiences Contribute to the Debate on Consciousness?

Peter Fenwick

Abstract The near death experiences (NDEs) is an altered state of consciousness, which has stereotyped content and emotional experience. Some features of the experience are trans-cultural and suggest either a similar brain mechanism or access to a transcendent reality. Individual features of the experience point more persuasively to transcendence than to simple limited brain mechanisms. Moreover there are, so far, no reductionist explanations which can account satisfactorily for some of the features of the NDE; the apparent "sightedness" in the blind during an NDE, the apparent acquisition after an NDE of psychic and spiritual gifts, together with accounts of healing occurring during an NDE, and the accounts of veridical experience during the resuscitation after a cardiac arrest. Although nonlocal mind would explain many of the NDE features, nonlocality is not yet accepted by mainstream neuroscience so there is a clear explanatory gap between reductionist materialistic explanations and those theories based on a wider understanding of mind suggested by the subjective experience of the NDEr. Only wider theories of mind would be likely candidates to bridge this gap.

8.1 Introduction

It is now almost 40 years since Raymond Moody published his seminal book *Life After Life* (1973). This book described near death experiences (NDEs) and brought these to the attention of both the lay and medical public. Many lay people

P. Fenwick (✉)
Kings College, London Institute of Psychiatry, London, UK

Department of Neuroscience, Southampton University, Hampshire, UK
e-mail: peter_fenwick@compuserve.com

A. Moreira-Almeida and F.S. Santos (eds.), *Exploring Frontiers of the Mind-Brain Relationship*, Mindfulness in Behavioral Health, DOI 10.1007/978-1-4614-0647-1_8, © Springer Science+Business Media, LLC 2012

interpreted these experiences as proof of an extension of consciousness beyond death (although Dr. Moody himself never originally claimed this), while those of the medical fraternity with a materialistic understanding of brain function rejected this interpretation and suggested that chemical changes within the brain were entirely responsible for the experiences. Over 13 million copies of the book have been sold, and it has been translated into 26 languages.

8.1.1   History of the NDE and Early Scientific Studies

Descriptions of similar experiences are found in myths and legends going back well over 2,000 years. The most ancient burial sites contain artefacts which suggest belief in the survival of some aspects of the human being after bodily death. Plato (427–347 BC) at the end of the *Republic* tells the story of a soldier who was thought to have died on the battlefield. He revived on his funeral pyre and described a journey out of his body to a place of judgement where souls were sent on to Heaven or to a place of punishment, according to the life they had lived on earth. Before reincarnation they were sent across a river where their experience of heaven was wiped from their memory. Ur himself was sent back to tell others what he had seen.

Noyes (1972) studied the NDE and delineated during the experience sequential stages of resistance, acceptance, and transcendence. Noyes and Slymen (1979), in the first attempt at statistical analysis of NDEs, categorised the experiences slightly differently, and found three general underlying factors: hyperalertness, with speeding up of thoughts and more acute vision and hearing; depersonalisation, with loss of emotion, and an altered sense of the passage of time, together with a feeling that the self was apart or detached from the body and felt strange or unreal. In about a quarter of their subjects they found a further, "mystical" factor, including feelings of great understanding, a sense of harmony and unity, feelings of joy and revelation.

Kenneth Ring, a psychologist at the University of Connecticut, was one of the first people to make a scientific analysis of these experiences. He listed the five features which appeared most commonly and which he found usually occurred in the same order – feelings of peace, the out of body experience, entering darkness, seeing the light and entering the light. This consistent pattern of events he called the "core" NDE. He went on to develop a more detailed scale which included meeting dead relatives, seeing beautiful colors or hearing music, encountering a being or presence and in some cases a "life review." He gave different weights to these features so that an NDE could be scored – the higher the score, the deeper the experience – WCEI, the weighted core experience index (Ring 1980).

Systemisation and standardisation between experimenters became possible when Greyson (1983a, b) applied the standard method of factor analysis to half of his data, and then tested out the weighted data on the second half. From this he was able to develop the Greyson scale, a standard assessment instrument which researchers now use to identify experience's of NDEs for research purposes. The NDE Scale consists of four sets of four questions, which identify cognitive, affective, paranormal, and transcendental NDE features.

8.1.2   *Phenomenology of the NDE*

The characteristics of the NDE in Western populations are well known. In those who suffer organic trauma, the experience is heralded by calm, which supercedes their pain. A small proportion then say they leave their body and view themselves, usually from a high vantage point. They may then go down a tunnel and enter the light at the end of the tunnel, where there is often a Being of light, with whom they will communicate. Often they may see dead relatives and sometimes a pastoral landscape with exotic flowers, heavenly music, etc. About 12% have a life review. They then progress to a border, which is also a barrier, which they realise is a point of no return. Here something or someone tells them, or they choose, to go back. There is no return journey; they usually "snap" back into their body. The features of an NDE do not necessarily occur in any particular order and not all may be present in a single experience.

Following a TV programme on NDEs in 1987 the author received 2,000 letters from people who had had NDEs. Questionnaires were sent to 500 of those correspondents who were thought to have the most comprehensive NDEs; finally over 450 replies were received. The importance of this data series is that 98% of them had had their experiences before they knew anything about NDEs and thus the series is uncontaminated by expectation, a sample which would be impossible to obtain now.

Calmness	80%
OBE	66%
Tunnel	82%
Being Of Light	50%
Pastoral Landscapes	76%
Meeting Friends/Relatives	38%
Life Review	12%
Barrier	24%
Decision To Return	72%
Transformation	72%
Hellish Experiences	4%

Although in our series people often described a tunnel like the tunnel in a spin dryer hose, many were much less specific and talked about a darkness through which they floated, sometimes with specks of light surrounding them which they interpreted as the light they saw in the distance reflected on the walls of a tunnel. Quite clearly, the description "tunnel" is a portmanteau term which suggests a small light shining in blackness, but it is the movement toward the light which tends to evoke the sensations of a tunnel (Fenwick and Fenwick 1995).

NDE have been found to occur in three completely different situations. First are the standard near-death situations where there is a clear organic component and brain function is significantly altered. Hypoxia and lowered or altered levels of consciousness are the hallmark of this group. For example, they can occur during a cardiac arrest, after a traumatic brain injury, during drowning, childbirth, feverish illnesses, or anaesthesia.

Table 8.1 Estimate of incidences in near death studies, N – population identified for study. n – sample of interviewees. (Permission granted by ABC- CLIO, LLC with thanks), From the handbook of near death experiences pp. 35

Study	Population	N	n	NDEs	Incidence	Criteria
Greyson 1986	Suicides	61	61	16	26%	WCEI
Greyson 2003a, b	Cardiac arrest	1595	116	27	23%	NDE scale
Greyson et al. 2006	Induced card.arr	52	52	0	0%	NDE scale
Milne 1995	Hemodynamic instability	86	42	6	14%	NDE scale
Orne 1995	Cardiac arrest	–	191	44	23%	NDE scale
Pacciolla 1996	Various resusc	125	64	24	38%	NDE scale
Parnia et al. 2001	Cardiac arrest	–	63	4	6%	NDE scale
Ring and Franklin 1981/2	Suicides	–	36	17	47%	WCEI
Schwaninger et al. 2002	Cardiac arrest	174	30	7	23%	NDE scale
Schoenbech and Hocutt 1991	Cardiac arrest	–	11	1	9%	NDE scale
Van Lommel et al. 2001	Cardiac arrest	344	344	62	18%	NDE interview

They can also occur in situations in which the person is not near death at all. They can, for example, be precipitated by extreme anxiety – the so-called "fear death" experience. There is some discussion in the literature whether these should be classified as NDEs or not (see below p.9). Finally there are the spontaneous, transcendent experiences, which have many of the features of the NDE, for example, the presence of love, light, and compassion, the intense animation of nature, sometimes the meeting with dead relatives, and often the impression that the individual is standing in the presence of some transcendent power.

The incidence of NDEs varies widely in different studies, depending on the population. A mean is about 20% – see Table 8.1. The most reliable figures for incidence come from the prospective studies of cardiac arrest, which suggest a figure of about 10% (Parnia et al. 2001; Greyson 2003a, b; van Lommel et al. 2001; Schwaninger et al. 2002).

The first prospective study was a small one, carried out at Southampton General Hospital by Dr. Parnia and his coworkers (Parnia et al. 2001). They asked a number of questions based on their knowledge of the retrospective accounts that were already published. Were the NDEs which were reported by cardiac arrest survivors similar to those which had already been reported in the retrospective literature? Could any clear organic features be likely causative factors in the genesis of the NDE in this very specific situation? Finally, was there any evidence that the OBE during cardiac arrest NDEs was veridical, i.e., when the subject claimed to be hovering just below the ceiling, looking down at the resuscitation situation, could he truly see or was what he reported simply a mind construct of what he would have expected to see? To check this, cards with pictures were placed just below the ceiling so that they would be visible to a patient who was truly seeing while out of his body. In all, there were 63 cardiac arrests and four reported NDEs. When rated on the Greyson scale these gave a score of over 7 and thus fulfilled the criteria for an NDE. The features of these NDEs was very similar to those already reported in retrospective studies, so the authors felt they could not draw a distinction between the two groups. No clear pathological features were noted in terms of length of resuscitation, drugs, or cardiac pathology, except those with NDEs had a significantly higher level of oxygen than those who did not. This was either a chance finding, or possibly, to be able report an NDE the brain had somehow been working more effectively, which it would probably have done with the increased oxygen levels.

More important, however, was the finding that none of those reporting NDEs had an OBE and thus none of the cards on the ceiling were viewed. The figures from the literature suggest that about a third of those who have NDEs during cardiac arrest also have OBEs, and using this data it was possible to power a future study. To be certain of collecting about 100 NDE experiences in which the subjects had OBEs, approximately 1,500 cardiac arrests would be needed. This would clearly be impossible in a single hospital study of any reasonable duration, and thus a multicenter trial was proposed. This trial was launched by Dr. Parnia in the autumn of 2008 at the United Nations at a conference on consciousness. The study is called AWaRE (awareness during resuscitation). To date 18 hospitals in the UK, two in the USA, one in France, one in Austria, and one possible hospital in Brazil are taking part in the study, and it is hoped more hospitals can be recruited. Approximately 60 cards are put up near to the ceilings in each hospital so that these could be viewed by patients who have a cardiac arrest NDE with an OBE. Data gathering has not yet started, but it is hoped there will be a result within 3 years.

Pim van Lommel's prospective cardiac arrest study of NDEs (2001) was comprehensive, well done, and extended our understanding of the relationship between cardiac arrests and NDEs and their consequences for the patient. Three hundred and forty-four patients in all had cardiac arrests and of these 62 (18%) did report an NDE. The depth of the NDE was classified using the WCEI scale. Very deep experiences were found in 2%, deep NDEs in 5%, moderately deep in a further 5%, some possible recollections in 6%, and 82% had no memory during the cardiac arrest. The characteristics of the NDE were similar to those of previous retrospective studies. A number of factors did not influence the occurrence of the NDE. Amongst these were the duration of the cardiac arrest and of unconsciousness, a complicated resuscitation, religion, etc. Factors that were found to increase the likelihood of an NDE were age (below 60) and previous NDEs. Importantly, those with lasting memory deficits were less likely to report an NDE. Of considerable interest was the finding that patients with deep NDEs were significantly more likely to die within 30 days. Van Lommel also did a comprehensive 2-year and 8-year follow up and found a number of factors on the life-change inventory, which were greater in those who had had an NDE: they were more loving and empathic, more involved with the family, had a stronger sense of the inner meaning of life and an interest in spirituality. Not surprisingly their fear of death decreased and their belief in a life after death increased.

Bruce Greyson studied 116 patients with cardiac arrests and found that 9.5% had reported an NDE with a score 6 or higher on the Greyson scale. A further 6% had memories during the arrest but did not score sufficiently highly to trigger an NDE on the Greyson scale. Again, this study found that the characteristics of the

NDE during a cardiac arrest were similar to the two previous prospective studies and also the retrospective studies. In summary, the author wrote "No one physiological or psychological model by itself explains all the common features of near death experiences...The paradoxical accounts of heightened, lucid awareness and logical thought processes during a period of impaired cerebral perfusion raises particularly perplexing questions for our current understanding of consciousness and its relation to brain function." He went on to suggest that NDEs "challenge the concept that consciousness is localised exclusively in the brain." (Greyson 2003a, b).

A more recent study of cardiac arrest and intensive care survivors has been carried out by Penny Sartori, an intensive care nurse, working in a hospital in Swansea. She found that only 1% of the 243 patients who survived the intensive care had NDEs, but this is not surprising as an intensive care unit has people who are extremely ill and others who come for only a short time just to be stabilised. She did however find 39 cardiac arrest patients, 18% of whom reported an NDE, and a further 5% who reported only an out of body experience without any NDE features, thus bringing the total to 23%. She was therefore able to ask whether the patients who said they left their bodies during the cardiac arrest were able to give a more accurate account of what happened during their resuscitation, than those who did not claim to have left their bodies or to have any memory of seeing the resuscitation. She asked both groups to describe what they thought had happened during their resuscitation and found that those who said they had seen the resuscitation were more accurate in their account of what had occurred than those who were simply guessing. This finding is important as it is the first prospective study which suggests that veridical information may indeed be obtained in some manner by someone who is deeply unconscious and who has none of the cerebral functions which would enable them either to see or to remember.

8.2 Outliers

Some aspects of the NDE – which I shall call outliers – do not fit neatly into a simple mechanistic and deterministic view of consciousness. They suggest that mind should be considered to have direct effects beyond the brain and thus has a non local component.

8.2.1 Wide Conscious States in the NDE Which are Supracultural

An excellent review of the cultural influences can be found in Kellehear (2009). He points out that in Indian, Tibet, Guan, and hunter gatherers of North and South America and Australia, there is no tunnel. The OBE is reported in all cultures apart from Australian and African hunter gatherers. Life review is seen mainly in Western and Asian cultures, but not in the Pacific areas or with hunter gatherers in the Americas, Australia, and Africa. Beings and other worlds are seen by all cultures. (Corrazza 2008) investigated Japanese NDEs and found that the tunnel was unusual and that often a river and a boatman were the method of transport to the next world.

This varied phenomenology, driven, it would seem, by cultural factors, must raise fundamental questions about the underlying physiology/pathology which gives rise to the features of the NDE. It has been suggested that if there was a mechanism which specifically produced the NDE, then NDEs would be similar across different cultures (Atwater 1988). However, this is a simplistic view, as there is abundant evidence that the content of a mental experience is in itself dependent on the culture in which it arises (Kellehear 2009). Pathological mental states vary widely between individuals and from culture to culture. For example, in toxic confusional states such as intensive care paranoid psychoses, although there may be an overriding paranoid feel to the experience, the actual details of the hallucinatory, delusional, and illusory experiences are determined by the individual and the situation in which the experience occurs (Page and Gough 2010). Thus variability is common and similarity less so.

In the case of the NDE the similarities are thus important and carry a significant weight over and beyond those of the differences. When stripped of cultural features, the phenomena which seem to be universal in the experiences are of "Beings" and the perception of another world. The important question for consciousness research is whether this world, which it seems can be accessed so easily in so many different ways, is fundamentally part of the structure of our consciousness in the same sense that the "real" world is. An alternative view is that the consciousness of these states, which is wider than everyday consciousness, suggests that there is a transcendent realm beyond that of the "ordinary" world to which the NDE has privileged access and which may be more than simply a matter of brain function. This suggests that entry into this world is an expanded form of consciousness.

8.2.2 Lack of a Single Explanatory Brain Mechanism

Should the NDE be described by its phenomenology only, or by the circumstances in which it arises? The variable conditions under which NDEs arise have led to a discussion of those circumstances which are essential for it to be classified as an NDE. For classification as an NDE, the experience must rate at least seven on the Greyson scale (see above). However, the scale takes no account of the circumstances under which the experience arises. As described earlier, NDEs arise from organic causes, e.g., severe brain injury or infection, but similar NDE phenomenology is also found in patients who are ill but not necessarily near death, and those who are not near death but are extremely anxious and "fear death." These psychological "fear death" experiences may occur in situations where the subject is terrified but not injured, such as just before a car accident.

Quite clearly specific mechanisms for the genesis of the NDE which are relevant to damaged brains will play no part in "fear death," transcendent and other NDE experiences, where there is no evidence of brain damage. Fox (2003) compared NDEs with experiences from the archives of the Religious Experience Research Centre at Lampeter UK, choosing those which had the same type of pattern as described by Moody (1973).

8.2.2.1 Moody's Criteria

1. A sense of ineffability after the experience
2. Hearing that they were dead
3. Feelings of peace and quiet
4. Hearing noises such as buzzing and wind-like sounds
5. A sensation of being out of the body
6. Passing through a tunnel
7. Meeting other individuals, often deceased relatives or friends
8. Encountering a Being of light
9. Having a life review
10. Reaching a border, crossing of which meant they would die
11. Finding they had returned to their physical body

After weighing up the evidence for either a similarity or dissimilarity of these wider transcendent experiences, he noted, that the classification of an NDE is "justified up to a point," and as emphasised by Sartori (2008, p. 12). We need to be aware that experiences which meet Moody's criteria can also occur in circumstances and in contexts in which there is no physical threat or danger.

Sartori reinforces Fox's suggested labeling that the experiences should be divided into crisis experiences and noncrisis experiences. Using this classification, Sartori goes on to note that the only NDE features absent from the noncrisis experiences were the border and the life review. It is thus highly unlikely that one single brain mechanism will explain every experience.

A third group of experiences, which are less cohesive and which again have the same phenomenology, are the transcendent experiences which may occur spontaneously during normal waking life. Some transcendent experiences will rate seven or over on the Greyson scale, but their phenomenology is looser than the other two groups described above. In this group of experiences, the subject is nowhere near death and thus the question of whether these are in the same category as NDEs becomes important. A further group are those experiences which again have the same phenomenology as the true NDEs but which occur when the subject is meditating, relaxing, or sometimes even dreaming (Fenwick and Fenwick 1995).

So it is clear that any theory which attributes the NDE to a particular organic process must be wrong, unless it applies only to a clearly defined subgroup, in which case the explanation is only partial. This is well reviewed by Greyson et al. (2009): "The multiple situations in which NDE phenomena arise, and the lack of a single, simple mechanistic brain-based explanatory system suggests that a wider consciousness framework should be considered." (Sect. 8.4 Models of Consciousness)

8.2.3   Mind Sight

A few people who have lost a primary sense, for example the sense of sight, have reported that they seem able to use this lost sense during an NDE. Adults blind from birth or under the age of five have no visual imagery (Berger et al. 1962). It would therefore be expected that the NDEs of visually deprived people would be significantly different from those of the sighted.

This idea was tested by Ring and Cooper (1999). They obtained 31 legally blind people who had had an NDE or OBE or both. Fourteen were blind from birth, 11 had lost their sight after the age of 5, and 6 were severely visually impaired. Eighty percent of their blind sample, including some who had been blind from birth, reported sighted experience during the NDE. The NDEs of the sample were described as containing the usual NDE components. The authors give a number of case histories, but one of their patients, described as never having had any visual experiences whatsoever, and had not even been able to differentiate light form dark, reportedly had two NDEs. In her second NDE, which followed an automobile accident, she claimed that she had left her body and from a point near the ceiling, had watched medical staff attending to her injuries, and described her floating body as being made of light. She then found herself rising through the roof of the hospital and moving toward the light. She came into an other-worldly pastoral realm where she saw trees, birds, and flowers, and dead friends and relatives. She met "Jesus" who was all light, had a life review, and then came back to her body. Many of the other patients' experiences were similar. Sixty-four percent of the congenitally blind people claimed visual experiences in their NDEs.

A few of their subjects claimed to have obtained visual information when out of their body. One claimed that when her heart had stopped she had seen two acquaintances standing in the hallway, and their presence was confirmed independently by the two people she claimed to have seen. The authors confirmed that NDEs do occur in the blind and are similar to NDEs in sighted people; they have visual impressions and the characteristics of the NDE were to some extent independent of the quality of the subjects' sight. This suggested to the authors that the visual experience could not be following the normal cerebral structures of sighted people. They describe it at one point as "eyeless vision."

The obvious criticism of this study is that blind people use sighted language – i.e., even though they may not see a TV, sitting and listening to it they describe as "watching" it. After examination the authors rejected this. They also discussed, but rejected, the theory that these were visual hallucinations, or an unconsciously generated fantasy that conformed to the subjects' view of what NDEs were like. Their preferred explanation is that the subjects depicted their experience in terms of visual impressions and that this was due to "mind sight," a transcendental state that allowed them to gather information and reformulate it in a way that they could comprehend.

It has been argued that the NDE experiences are simply dreams. But congenitally blind people and those who lost their sight before the age of 5 have no visual imagery in their dreams. In those blinded at a later age, the visual imagery fades with time. There has been no prospective study to support the suggestion that the blind sighted experiences during an NDE are veridical, but the retrospective accounts from the above authors provide some evidence that they could be. Clearly more studies are required to clarify the issues raised.

If we accept the data from the limited study of Ring and Cooper, then the NDE in the congenitally blind seems to open up a new "sensory pathway," which gathers information without a sensory organ – the authors' "mind sight." This is similar to remote viewing, in which a normally sighted, trained subject is able to gather visual information about a remote target, Schwartz has collected over 80 scientific papers written on this subject, and so there is a scientific data base from which to judge this phenomenon (Schwartz 2011, web site). It seems unlikely that there are two different mechanisms for this similar phenomenon. However, I am unaware of any prospective studies carried out on remote viewing in the blind, so it is not possible to know whether the mechanism is indeed the same. The significance for consciousness research is again clear as it suggests, if the conclusions of Ring and Cooper's study is correct, then that mind must extend nonlocally and obtain information beyond the range of the senses.

8.2.4   Transformation After an NDE

It is not surprising that subjects who have an NDE following a life-threatening event should be transformed by the experience. Any life-threatening event does of course produce change, but the data point toward NDErs being more positively affected by the change than those who had similar life-threatening experiences but had not had an NDE. Greyson (2003a, b) reported on 272 patients who had had a "brush with death," 61 of whom (22%) had had NDEs. They were found to be less psychologically disturbed after the experience than those who had not had NDEs.

Two studies from patients with cardiac arrest are sufficiently comprehensive to make this point. van Lommel et al. (2001) rated the quality of life of subjects postevent, using the Life Change Questionnaire. Changes in social and religious attitude were positively affected. Fear of death was reduced, and search for personal meaning, interest in self-understanding, and appreciation of ordinary things were enhanced (see p. 7). A smaller study by Schwaninger et al. (2002), using the same questionnaire, found essentially the same results. For a full review see Noyes et al. (2009).

A number of factors during and preceding the experience are thought to influence the development, significance, and range of after-effects, including the circumstances, closeness to death, and – particularly – depth of the experience (Greyson and Stevenson 1980; Schwaninger et al. 2002; van Lommel et al. 2001). There is also a relationship between specific elements of the NDE and its after-effects: the out of body experience, for example, suggests to the experiencers the possibility of continuation of consciousness after death and is a model for the soul leaving the body (amongst others Tiberi 1993). Personality traits may also be important. A number of studies have looked at both pathological and normal psychological characteristics. Essentially, NDErs are indistinguishable from a number of comparison groups as regards their mental health. With regard to nonpathological characteristics, there is some suggestion that the capacity for psychological absorption or fantasy proneness increases the chance of having an NDE, but the evidence is not compelling (Greyson 2000). For a review see Holden et al. (2009, pp. 126–130). However, there is a long history of similar transformation of people who have had a deep transcendent experience, even if it does not fit neatly into the NDE phenomenology (Fox 2003). This suggests that any strong spiritual experience can produce positive spiritual change.

What is of more relevance for this article is that there is also evidence that NDErs report an increase in subsequent paranormal experiences, together with an alteration in perception and consciousness. A number of authors report that following the NDE, there are occurrences of precognition, intuition, guidance, clairvoyance, telepathy, out-of-body experience, and healing ability. Some reported that together with the healing ability, there was perception of other people's auras, and some even claimed to be able to contact spirits. It has been suggested that this is equivalent to the development of psychic awakening. Sutherland (1989) describes the ability of her NDE subjects to come to terms with clairvoyant or precognitive experiences, whereas out of body or telepathic experiences they found disturbing and so tended to suppress them. No study that I know has tested objectively the paranormal claims among NDErs, so their validity remains subjective.

There are also accounts of changes in states of consciousness and in our sample the occurrence of Kundalini-like phenomena (Kundalini is a yoga term indicating an energy source at the base of the spine, which rises through the body producing different sensations and leading to wisdom) reported following the NDE. These usually took the form of shocks or jolts radiating through the body, and the person's knowledge that they were on a progressive path which would be likely to lead to an expansion of consciousness (Fenwick and Fenwick 1995; Noyes et al. 2009).

There are also claims of healing by the experience itself. The most high profile, and controversial, of these is the experience described by Mellon Thomas Benedict, who claims that he was "dead" for 1 h during which he had an NDE, after which he maintains that he was cured of a brain cancer (http://www.mellen-thomas.com/quotes.html). In a good prospective study of intensive care patients and cardiac arrests in the cardiac care unit, Sartori (2008, p. 238) found that 13% of the NDErs in her study reported spontaneous healing following their NDE, whereas none of the control patients did so. One such patient came into hospital after a road traffic accident with chest trauma, a liver tear, and a fractured right humerus. His clinical state was too severe for him to be operated on immediately. When he was finally discharged to the ward from ITU and the surgeon went to fix the patient's arm, he found that it had already healed. This case was considered by the staff on the ITU to be an unusual event. A second patient had had a contracture of his hand for 60 years, following a mild cerebral palsy and a spastic hemiparesis of his right hand, well documented in the patient's notes and confirmed by his wife and sister. He had an OBE during his NDE in which he found he was able to move his hand normally. Afterwards he found that he was able to move his previously contracted hand normally. This was investigated by a physiotherapy team on the ward who reported that this could not have occurred without special surgery to release the contracture. Thus, the healing of his hand has to be recognised as an NDE healing.

The personality changes, the acquisition of "special gifts," and spontaneous healing unexplained by a direct physical intervention, raise questions about the relationship of consciousness to the body and to the mind which would need to be taken into account in any theory of consciousness. It must be recognised that spontaneous healings are well reported in the literature (O'Regan and Hirschberg 1993) but these are usually seen as interventions by the immune system, leading for example to the regression of cancers. Clearly this mechanism cannot explain all cases and thus the NDE should be seen as a good model for investigating the wider effects of these experiences. In the esoteric and meditation literature, there are frequent references to the development of special powers (the siddhis) after spiritual training or deep spiritual experiences. It would be reasonable to accept that the NDE would count as such a deep experience and thus might also results in changes in subjective experiences.

Although a few people say they have acquired psychic gifts such as telepathic powers, clairvoyance or the ability to heal after an NDE, no studies so far have been carried out to investigate the veridical nature of these claims (Greyson 1983b; Sutherland 1989). Further investigation of these phenomena would be helpful, as if they were validated, a simple mechanistic description of mind would not explain the data.

8.2.5   Claims of Veridical Experience

Veridical perception is well reviewed by Holden (Holden et al. 2009, p. 185), who has found in the literature 107 possible cases of nonphysical veridical perception. She defines it as "any perception, visual, auditory, kinaesthetic, olfactory, and so on – that a person reports having experienced" during an NDE, which is later confirmed by an independent witness and cannot be explained by the normal senses. She describes two kinds of veridical perception, that occurring in the physical domain, e.g., a member of the family speaking in a different room, while the NDEr was ostensibly under an anaesthetic, or in the mental (spiritual) domain, e.g., when the NDEr meets, in his NDE, a relative who has in fact died, though the NDEr does not know this. Out of body perceptions which are claimed to occur at the onset of a cardiac arrest are particularly important for our understanding of consciousness, because they open up the possibility of confirming whether these are indeed veridical. See AWARE study above, Sect. 8.1.2.

Veridical data raises direct questions about the nature of mind. It suggests the possibility that the senses can gather information remotely, i.e., that during an NDE the mind is extended. Examples from our own series include the account given to her parents by a 5-year-old girl after a tonsillectomy operation. She described how the doctors "had funny scissors with long handles and they go snip-snip in your throat." Was this an OBE? It seems highly unlikely that anyone would have tried to prepare a 5-year-old for a tonsillectomy by describing the operation to her in quite such graphic terms, and equally unlikely that she had managed to see the surgical instruments before the operation.

Another convincing account was given by a man who suffered two cardiac arrests while he was in intensive care.

"It was as though I was standing on the wall of the ICU defying gravity and looking down at my own body. I was shocked at what an ugly corpse I was. I was naked and the nurse was taking down a drip out of my ankle. I vividly remember how purple my face was and how blank my forehead seemed. I appeared to have a black triangle from my hairline to my nose. My wife later confirmed that that was how I looked when she was allowed in to see me."

One of the most frequently repeated stories is that of Kimberley Clark, who during her NDE left her body and saw a shoe on a window ledge. Having recovered, she was able to check the window ledge where she did indeed find the shoe. This story has been criticised by Dr. Sue Blackmore, who tried to find Kimberley Clark but was unable to do so and thus felt that the story lacked confirmation. However, I have met Kimberley Clark and she confirms the story. There are already well conducted studies, which suggest that remote viewing occurs in other circumstances (see Sect. 8.2.3). This is further evidence that extended mind may need to be included in any model of consciousness.

8.2.6 Prospective Studies of Cardiac Arrest

Two important outliers occur in NDEs associated with cardiac arrest. Ten percent of patients with cardiac arrests have an NDE (Parnia et al. 2001; van Lommel et al. 2001; Schwaninger et al. 2002; Greyson 2003a, b). Of these, about 30% will have an out of body experience. Cardiac arrest NDEs with out of body experiences are of interest as the experiencers report that while they are unconscious and being resuscitated their sense of subjectivity leaves the body, goes to the ceiling and watches the resuscitation process. However, at that time the brain is nonfunctional, so according to reductionist views of mind, both perception and memory of that perception should be impossible.

There are now one or two possible prospective cases of veridical perception during a cardiac arrest. The most widely discussed, though often disputed, prospective case is that of Pamela Reynolds (Sabom 1998), who was diagnosed with an aneurysm deep within the brain. In order that the surgeons could clip the aneurysm satisfactorily, it was necessary to cool her, stop her heart, empty the blood from her brain, and then carry out the surgical procedure. During this process she had an OBE in which she was able to identify the cranial saw and a number of other instruments and the operating room personnel. She saw and heard an attempt to insert a catheter unsuccessfully into her right groin, and then successfully into her left groin. Following the OBE, she had an NDE experience in which she saw dead loved ones and was accompanied by her dead uncle who returned her to her body.

Sartori (2008) describes several cases of out of body experience during unconsciousness, some with veridical perception, which support the previous retrospective studies. One patient was making a good recovery following emergency surgery when his blood pressure suddenly dropped and he lost consciousness. He was transferred back to bed while deeply unconscious and not responding to stimuli and remained deeply unconscious for 30 min, not regaining full consciousness for 4 h. During the period of unconsciousness, he had an OBE in which he correctly reported that one of the doctors shone a torch in his eye, that he watched the intensive care nurse clean his mouth and described the instruments she used. He also reported that a physiotherapist put her head around the curtain to find out how he was. All these reports were accurate.

In a cardiac arrest, the three major signs of clinical death are present: no cardiac output, no respiration, and an absence of brainstem reflexes. Yet if NDEers perception at this time is indeed veridical, it is then during this period of the arrest that the NDE experience occurs. It is often argued that even though the clinical signs of brain death are present, the brain state is reversible and thus there may be some small areas of brain which are still functioning and can construct the experience. However, the features of the experience are so wide, containing emotion, visual and auditory, tactile and sometimes olfactory sensations, that no single small brain area would be able to generate the experience. It is also important to note that memory is always affected by severe cerebral disruption, and thus a straightforward explanation using normal cerebral processes is not feasible. Another argument is that the NDE occurs during the onset of unconsciousness, but this is always rapid and thus the experience cannot occur at that time. Eleven seconds after a cardiac arrest, electrical silence supervenes and the brain is totally dysfunctional (Koenig et al. 2006). It has also been suggested that it occurs during the recovery period, but recovery from a cardiac arrest is always confusional and these experiences are lucid. When resuscitation is started, it is unusual for the blood pressure to increase to more than 30 mm Hg, which is not sufficient for the cerebral circulation to be reestablished. The brain does not become functional again until after the heart has restarted. It is difficult to see how a brain which is severely impaired by anoxia and hypercapnoea with distorted electrical transmission would be able to build such clear, lucid, and comprehensive experience as the NDE (Howard et al. 2011).

Any firm conclusion drawn from these experiences must depend on the precise timing of the experience, and thus on the veridical nature of the perception reported. The AWARE study (awareness during resuscitation) is a prospective study, which is setting out to gather cases of NDE out of body experiences during cardiac arrest. Sixty boards with information on have been put up above the beds in the hospitals resuscitation area, in each of over 18 hospitals in the UK and others in Europe and

America, with the aim of collecting over 100 cases of out of body experiences during a cardiac arrest, and testing to see whether the boards have been viewed. This study, run by Dr. Sam Parnia, is the first comprehensive study setting out to examine the nature of consciousness during cardiac arrest. If it can be shown that veridical perception is possible, then the relationship between consciousness and brain function will have to be rethought. (Parnia 2007).

8.3   The Significance of the Outliers

8.3.1   Wide Supracultural Conscious States

The fact that many different cultures have similar content to the NDEs suggests the possibility of a common brain programme which occurs at this time. However, a transcendent realm which may be accessed by the experience could also be important and this would then suggest that consciousness is more than simply a matter of brain function.

8.3.2   Lack of a Single Explanatory Brain Mechanism

No single, simple mechanistic brain-based explanatory system has been found for the many different situations in which NDEs can occur. This suggests that a wider consciousness framework common to all conditions should be considered.

8.3.3   Transformation

The very deep spiritual experience of the NDE suggests that the expression of consciousness by the individual is changed and a number of new spiritual qualities are added.

8.3.4   Veridical Experience and Data from Cardiac Arrest Studies

If veridical data acquired in an NDE can be confirmed, it would indicate that the mind can gather information remotely. This would suggest that during an NDE mind/consciousness may be extended beyond the brain. In this case models based solely on the physical properties of the brain, which limit consciousness to the brain, may need to be widened.

8.4   A Simple Guide to Models of Consciousness

There is no accepted brain mechanism by which subjective experience, or conscious-ness, can arise from the objective brain structures. Brain neuroscience has advanced to the point where the functions of major areas within the brain are known. Systems of activity during the alert state, resting (the default state), and during sleep and dreaming are all well mapped. Neuroimaging has shown numerous correlates between function and brain areas but it is important to realise that these findings are only correlates. If we feel anger, the left amygdala may well show increased activ-ity, but we do not know how this increased activity translates into the emotion we feel (Adolphs 2008). The gap between objective measurement of the brain and sub-jective experience has been well shown by the philosopher Nagel (1974) using the example of a bat. However, much we know about the neurophysiological function-ing and structure of a bat's brain, its radars and sensory mechanisms, we will never know what it is like *to be* a bat. There is still no answer to Chalmers' "hard prob-lem" (Chalmers 1995, 2010) in that even with our profound objective understanding of brain mechanisms, it is still impossible to get from those to subjective experience. What is it about the NDE that might extend our understanding of the relationship between brain function and consciousness?

Models of consciousness have now become highly sophisticated and are beyond the scope of this article. For a review see *The Handbook of Consciousness* (Velmans and Schneider 2006). However, with our current understanding, there are three basic models.

8.4.1   *Reductionist Models*

The reductionist models argue that the material world is all there is, and material explanations are the only valid ones. The difficulty with this model is that mind and consciousness are excluded as causal agents. The usual way round this is to use the concept of "emergence," which states that it is impossible to know what new proper-ties will arise from a combination of material substances. Thus it is logical to argue that consciousness itself may emerge with the right configuration of material ele-ments. The usual example given for emergence is the combination of hydrogen and oxygen, both gases, to produce the entirely new substance, water. This is clearly a flawed argument, because both oxygen and hydrogen, when cooled or compressed sufficiently, turn to liquid. Thus liquidity is within their nature. Others have argued that emergence is truly *de novo,* and there is no way of predicting it. If that is so, then the concept of emergence has no scientific value, that is, it is never possible to predict what will emerge, and so find a causative chain of explanation based only in matter. This idea of a causative chain was dealt with by de Haene and Naccache (2001), when they argued that the cognitive neuroscience of consciousness sets out to determine whether there is a systematic form of information processing and a specific and reproducible class of neuronal activation patterns that systematically (and very importantly) are able to distinguish mental states the subjects label as conscious, from other states which are not. Molyneux (2010) argues that this is purely a referencing problem and not soluble as we lack a procedure for definitely ruling out neurological and theoretical contenders. He argues that neural correlates should not be our only goal, but we must come to understand how mechanisms of a certain sort must produce consciousness. He feels that this will bridge the explanatory gap, but I am not certain that this moves the argument very much further forward.

8.4.2   Consciousness as Primary

The next group of models postulates that consciousness is the basic structure of the universe and that only those models which postulate consciousness as primary can successfully explain the problems facing neuroscience today. Like the first model, this has an inherent flaw – the problem of the explanatory gap. In the reductionist models, a small miracle is needed for consciousness to arise from physical processes; in the consciousness models, the miracle relates to the arising of matter from consciousness itself.

An interesting idea is the possibility of transferring consciousness between physical structures and so extending them to possible receptacles for human consciousness at death. An excellent compendium from the *Journal of Consciousness Studies 2010*, entitled *The Singularity*, considers what would happen if machines became more intelligent than humans and were able to download consciousness in to a mechanical structure. This has many references to the film *The Matrix*. It gives the authors the opportunity to look more closely at the relationship between the neuronal systems of the brain, the physical systems of biology, the evolutionary systems of biological, social, and cultural development and finally the interlinking of these ideas with universal consciousness and the way this may link up with these physical processes.

Chalmers, in an introduction to *The Singularity* asks whether artificial intelligence, (AI) or super artificial intelligence (AI+) can ever reach the point at which brain information can be down-loaded into a new physical structure which will hold our consciousness, so that with the death of the brain we can remain "alive." He argues, as he has before, that the physical structure which carries out all the behaviors of a conscious person is not necessarily conscious, and argues here that there is a difference between biological and functional theories. He says it "is crucial to the practical question of whether (or) not we should upload. If biological theories are correct, uploads cannot be conscious, so we cannot survive consciously in uploaded form. If functionalist theories are correct, uploads almost certainly can be conscious and this obstacle to up-loading is removed… ….. It is true that we have no idea how a non-biological system, such as a silicon computational system, could be conscious. But the fact is that we also have no idea how a biological system such as the neural system could be conscious. The gap is just as wide in both cases" (Chalmers 2010).

8.4.3 Dualistic Models

The third group, dualistic models, postulates that both mind (consciousness) and matter exist as independent entities but are linked closely together. The complexity of the models essentially relates to the degree of linking which is allowed. They can be absolutely linked so that every material element right down to the fundamental particles has consciousness inherent in it. Thus, on one view the particle is material and on another it is an element of mind. These models are attractive as they overcome the difficulty of the emergence of mind from a physical entity or vice versa. But there remains the difficulty of explaining how mind and brain interact and the nature of the mind–brain interface. Simply stating that brain looked at one way is material and looked at another way is mind and that they are closely interlinked still leaves us with the problem of explaining, with our current understanding of neuroscience, how the objective and subjective sides emerge together. Dualistic models echo the current vogue of studying both Eastern and Western psychology, Eastern with meditation and the study of qualitative changes in mental state, and Western which bases its understanding on the behavior and knowledge of physiological processes relating to brain function, and more recently, the recognition that sociobiological factors are important.

Lockley has taken these ideas further, based on exploring Gebser's classic work on consciousness and Goethian ideas relating to the development of biological systems, embedded as they are in an overriding field of consciousness. He argues that these theories, developed in evolving structures, are responsive to a much wider set of influences which condition the way they evolve and the patterns they take. He argues for a cosmic, pranic energy, which conditions the evolutionary forms and is closely linked to universal consciousness itself, thus binding together consciousness and matter into a single system (Lockley 2010).

However, when considering dualistic theories, the most important contribution must come from those who have worked intimately with the human brain and its stimulation during surgical procedures. Wilder Penfield (1970) said "For myself, after a professional lifetime spent in trying to discover how the brain accounts for the mind, it comes as a surprise now to discover, during this final examination of the evidence, that the dualist hypothesis (separation of mind and brain) seems the more reasonable of the two possible explanations…Mind comes into action and goes out of action with the highest brain-mechanism, it is true. But the mind has energy. The form of that energy is different from that of neuronal potentials that travel the axon pathways. There I must leave it."

8.4.4 Field Theories

Field theories have a long history, dating back to Plato who described the idea of a transcendent reality in the *Republic*. He suggested that the life we ordinarily live in the cave (this world) is a reflection of shadows cast, by true transcendent reality which exists beyond the cave. This idea of ultimate truth existing as the creative background to reality has become the basis for field theories. These theories essentially argue that the wide consciousness of the cosmos underpins all conscious experiences but as it is so wide it is of necessity filtered by the brain. Emmanuel Kant is famous for his distinction between "the thing in itself," i.e., that which is outside the realm of the senses, and "the thing for us," which we can know (*Critique of Pure Reason* 1781). He describes this position as follows: "The body would thus be not the cause of our thinking, but merely a condition restrictive thereof. Though essential to our sensuous and animal consciousness, it may be regarded as an imposition to our pure spiritual life." William James (1842–1910) in his Gifford lectures of 1901–1902, which were later published as *The Varieties of Religious Experience* (1902), also suggested that consciousness could be considered to be a field. James suggested that while ordinarily the brain acts as a reducing agent so that our cognitive perceptions are limited, in transcendent experiences a change in the "filter mechanism" of the brain allows the field of mind to be extended into the transcendent. Field theories are not dualistic theories. They do not argue that brain is one substance and mind another. The field theory argues that individual consciousness is part of universal consciousness and is filtered down so that it can be restricted in its use. Brain substance, rather than being different from universal consciousness, is just another manifestation of it, so you no longer have the difficulty of two substances, which cannot communicate with each other.

Wide transcendent states are experienced to some extent by 30% of the population, and more profoundly in 10% (Hay 1990). In his seminal book, *The Spiritual Nature of Man,* Sir Alastair Hardy describes these states and the Alastair Hardy Centre in the UK has been collecting descriptions of them from the general public (Hardy 1979). An example is given by Nona Coxhead (1985) "…Suddenly the entire room was filled with a great golden light. The whole world was filled with nothing but light. There was nothing anywhere except this effulgent light and my own small kernel of the self. …Extraordinary intuitive insights flashed across my mind. I seemed to comprehend the nature of things. …Neither time nor space existed on this plane…." These transcendent experiences have many of the features of the NDE; the experience is always described as more "real" than that of the everyday universe, which is perceived as alive and conscious.

If we consider that there is a transcendent reality over and above that of the ordinary everyday world, that mind is extended, and consciousness is not limited to the brain, then many of the outliers of the NDE which are described above fall easily into place. The presence of a transcendent world which is supra-cultural and the ease with which it can be accessed fits well. An extended mind, again shown by the NDE, will also fit well into this model. Provided the transcendent is given priority in a causal sense, then healing and transformation of the personality with supra-sensory gifts after an NDE could also be accounted for. The major benefit of these theories is that they are nonlocal. The transcendent mind is universal and the individual mind has access to it. Thus if this is so it is not surprising that the NDE points to the lack of theories relating to consciousness within the brain and our inability to find the neural basis of consciousness, which is of course much wider and includes a transcendent component. Mindsight and veridical experience are also more easily explained using a holistic transcendent framework. But importantly, a field transcendent theory postulates a universe, which is much closer to that experienced by the NDErs than the one that can be accounted for by the shuffling about of our current, limited reductionist neuronal correlate theories.

References

Adolphs, R. (2008). Fear, faces, and the human amygdala. *Current Opinion in Neurobiology, 18*(2), 166–172.

Atwater, P. (1988). *Coming back to life: The after-effects of the near-death experience*. New York, Valentine: Citadel.

Berger, R. J., Olley, P., & Oswald, I. (1962). The EEG, eye movement and dreams of the blind. *Quarterly Journal of Experimental Psychology, 14*, 183–186.

Bibliography of Scientific Remote Viewing Research Papers

Chalmers, D. J. (1995). Facing up to the problem of consciousness. *Journal of Consciousness Studies, 2*, 200–219.

Chalmers, D. J. (2010). The singularity of philosophical analysis. *Journal of Consciousness Studies, 17*(9–10), 7–65.

Compiled by Vernon Neppe, MD, PhD and Stephan A. Schwartz

Corrazza, O. (2008). *Near death experiences, exploring the mind-body connection* (pp. 102–117). Oxon, UK: Routledge. Chapter 5.

Coxhead, N. (1985). *The relevance of bliss: A contemporary exploration of mystical experience* (p. 35). London: Wildwood House.

Dehaene, S., & Naccache, L. (2001). Towards a cognitive neuroscience of consciousness: Basic evidence and a workspace framework. *Cognition, 79*, 1–37.

Fenwick, P., & Fenwick, E. (1995). *The truth in the light: An investigation of over 300 near death experiences*. London: Hodder Headline.

Fox, M. (2003). *Religion, spirituality and the near death experience*. London and New York: Routledge.

Greyson, B., & Stevenson, I. (1980). The phenomenology of near-death experiences. *American Journal of Psychiatry. 137*, 1193–1196

Greyson, B. (1983a). The near death experience scale: Construction, reliability, and validity. *The Journal of Nervous and Mental Disease, 171*, 369–375.

Greyson, G. (1983b). Increase in psychic phenomen following near death experiences. *Theta, 11*, 26–29.

Greyson, B. (2000). Near death experiences. In E. Cardena, S. Lynn, & S. Krippner (Eds.), *Varieties of anomalous experience: Examining the scientific evidence* (pp. 315–352). Washington, DC: American Psychological Association.

Greyson, B. (2003a). Near-death experiences in a psychiatric outpatient clinic population. *Psychiatric Services, 54*(12), 1649–1651.

Greyson, B. (2003b). Incidence and correlates of near-death experiences in a cardiac care unit. *General Hospital Psychiatry, 25*, 269–276.

Greyson, B., & James, D. (2009). The Handbook of Near-Death Experiences. Praeger Publishers. Oxford. England.

Greyson, B., Kelly, E. W., & Kelly, E. F. (2009). Explanatory Models for the NDE Experience. In B. Greyson, J. M. Holden, & D. James (Eds.), *The handbook of near death experiences: Thirty years of investigation*. Santa Barbara, CA: ABC-CLIO, LLC. Chapter 10.

Hardy, Sir Alastair 1979 The Spiritual Nature of Man. Oxford University Press.

Hay, D. (1990). *Religious experience today: Studying the facts*. London: Cassell.

Holden, J. (2009). In J. Holden, B. Greyson, & D. James (Eds.), *The handbook of near-death experiences* (p. 185). Oxford, UK: Praeger Publishers. Chapter 9.

Holden, J., Long, J., & MacLurg, J. (2009). In J. Holden, B. Greyson, & D. James (Eds.), *The handbook of near-death experiences*. Oxford, UK: Praeger Publishers. Chapter 6.

Howard, R. S., Holmes, P. A., & Koutroumanidis, M. A. (2011). Hypoxic-ischaemic brain injury. *Practical Neurology, 11*(1), 4–18.

Kellehear, A. (2009). Census of non-Western near death experiences to 2005: observations and critical reflections. In B. Greyson, J. M. Holden, & D. James (Eds.), *The handbook of near death experiences: Thirty years of investigation* (p. 135). Santa Barbara, CA: ABC-CLIO, LLC. Chapter 7.

Koenig, M. A., Kaplan, P. W., & Thakor, N. V. (2006). Clinical neurophysiologic monitoring and brain injury from cardiac arrest. *Neurologic Clinics, 24*(1), 89–106.

Lockley, M. G. (2010). The evolutionary dynamics of consciousness: An integration of Estern and Western holistic paradigms. *Journal of Consciousness Studies., 17*(9–10), 66–116.

Molyneux, B. (2010). Why the neural correlates of consciousness cannot be found. *Journal of Consciousness Studies, 17*(9–10), 168–188.

Moody, R. (1973). *Life after life*. Atlanta, Georgia: Mockingbird.

Nagel, T. (1974). *What is it like to be a bat?*.: Philosophical Review.

Neppe, V., & Schwartz, S. A. Bibliography of Scientific Remote Viewing Research Papers

Noyes, R. (1972). The experience of dying. *Psychiatry, 35*, 174–184.

Noyes, R., Fenwick, P., Holden, J., & Christian, S. (2009). After-effects of pleasurable western adult near-death experiences. In B. Greyson, J. M. Holden, & D. James (Eds.), *The handbook of near death experiences: Thirty years of investigation* (p. 51). Santa Barbara, CA: ABC-CLIO, LLC. Chapter 3.

Noyes, R., & Slymen, D. (1979). The subjective response to life- threatening danger. *Omega, 9*, 313–321.

O'Regan, B., & Hirschberg, C. (1993). Spontaneous remission: An annotated bibliography. *Institute of Noetic Science*.

Page, V., & Gough, K. (2010). Management of delirium in the intensive care unit. *British Journal of Hospital Medicine (Lond), 71*(7), 372–376.

Parnia, S. (2007). Do reports of consciousness during cardiac arrest hold the key to discovering the nature of consciousness? *Medical Hypotheses, 69*(4), 933–937.

Parnia, S., Waller, D. G., Yeates, R., & Fenwick, P. (2001). AA qualitative and quantitative study of the incidence, features and aetiology of near death experiences in cardiac arrest survivors. *Resuscitation, 48*, 149–156.

Ring, K. (1980). *Life at death: A scientific investigation of the near death experience*. New York: Coward, McCann, and Georghegan.

Ring, K. (1984). *Heading towards omega: In search of the meaning of the near death experience*. New York: William Morrow.

Ring, K., & Cooper, S. (1999). *Mindsight: Near death and out of body experiences in the blind* (William James Center for Consciousness Studies). Palo Alto, CA: Institute of Transpersonal Psychology.

Sabom, M. (1998). *Light and Death: One doctor's fascinating account of near-death experiences*. Grand Rapids, MI: Zondervan

Sartori, P. (2008). *Near death experiences of hospitalised intensive care patients: A five year clinical study* (p. 238). Lampeter: Edwin Mellon Press.

Schwaninger, J., Eisenberg, P., Schechtman, K., & Weiss, A. (2002). A prospective analysis of near death experiences in cardiac arrest patients. *Journal of Near Death Studies, 20*(4), 215–232.

Schwartz, S. (2011). http://www.stephanaschwartz.com/category/papers-and-research-reports/selected-bibliographies-of-nonlocal-research/

Sutherland, C. (1989). Psychic phenomena following near death experiences: An Australian study. *Journal of Near-Death Studies., 8*, 93–102.

Tiberi, E. (1993). Extra-somatic emotions. *Journal of Near-Death Studies, 11*, 149–170.

Van Lommel, P., van Wees, R., Myers, V., & Elfferich, I. (2001). Near death experiences in survivors of cardiac arrest: A prospective study in the Netherlands. *Lancet, 358*, 2039–2045.

Velmans, M., & Schneider, S. (2006). *The Blackwell companion to consciousness.*: Wiley-Blackwell.

Chapter 9
Death, End of Life Experiences, and Their Theoretical and Clinical Implications for the Mind–Brain Relationship

Franklin Santana Santos and Peter Fenwick

Abstract The attitude towards death and dying depends on the culture. In prehistoric times grave artifacts suggest a belief in the continuation of life. This belief in an afterlife has continued through different cultures and societies to the present day. The fear of death seems to have grown in parallel with those religions which have promised judgment at the time of death. In our modern Western secular society death is regarded as a medical failure, the rituals which used to attend it have largely been abandoned, and life is prolonged so that death has lost all dignity. It is now beginning to be recognised that dying may not be a simple switching off, but a process leading to death and the gradual dissolution of consciousness. This dissolution seems to involve experiences for the dying which are spiritual and important for them. A number of these phenomena raise the possibility that consciousness may not be limited to the brain, but extend beyond it. Fortunately, palliative care is now taught in medical schools, and treatment of the dying is now recognised to be as important as treatment for the living. This article looks at the history of death, the significance of the dying process for consciousness research, and the education needed for carers of the dying.

F.S. Santos (✉)
Institute of Psychiatry, School of Medicine, Sao Paulo University, 05403-010, São Paulo, Brazil
e-mail: franklin@saudeeducacao.com.br

P. Fenwick
Kings College London Institute of Psychiatry, London, UK

Department of Neuroscience, Southampton University, London, UK
e-mail: peter_fenwick@compuserve.com

A. Moreira-Almeida and F.S. Santos (eds.), *Exploring Frontiers of the Mind-Brain Relationship*, Mindfulness in Behavioral Health, DOI 10.1007/978-1-4614-0647-1_9, © Springer Science+Business Media, LLC 2012

9.1 Introduction

The historian David Stannard (1975, 1977) points out that in societies in which the individual is unique, important, and inimitable, death is not ignored, but is marked by a sort of collective mourning over the social loss of one of its members. On the contrary, in societies in which people feel that the loss of one individual outside one's immediate circle has little effect on society, death receives scant, if any, attention. The first step in paying adequate attention to death is to recognize that in avoiding or denying it, we are denying an integral part of human life.

The study of death and the question of survival concern issues rooted at the very core of human life. Thus, the person who wishes to augment his knowledge of consciousness, mind–brain relationship, death and the act of dying is embarking upon an exploration which is nothing more than a trip to discover oneself. The study of death and its implications for the mind–brain relationship is a journey to the inner self and knowing oneself is only possible through an interdisciplinary approach combining medicine, human, and social sciences (Santos and Incontri 2008; Santos 2009a, b, 2010a, b, c). Death deals with the most fundamental, intriguing, and challenging question, the one most disquieting for humanity: the question of what comes after life. Neither religion, philosophy, nor science has been able to answer this satisfactorily: death remains a somber figure whose presence is only tenuously perceived, relegated to the periphery of our lives.

Today, more than at other time in the history of humanity, we would like, if not to deny death, at least to control it, by means of the advances in the biological sciences over the last two centuries (Ariès 2000a, b; Feifel 1959, 1977). The advance of cloning techniques, for example, even opens up the possibility that we may acquire a new body and a new mind, suggesting the possibility of immortality.

In this brief reflection on death and dying and its consequences to the mind–brain relationship: we would like to provide some observations that enable deeper reflection on a subject which should no longer be regarded as of secondary importance but something of fundamental interest to both the academic and lay public.

9.2 Death in Prehistoric and Ancient Times

Death is a universal human experience. Dying and death itself are more than biological phenomena: they have religious, social, philosophical, anthropological, spiritual, and academic dimensions (Santos and Incontri 2008; Santos 2009a, b, 2010a, b, c). Questions about the meaning of death and what happens when we die have been central preoccupations in every culture since time immemorial.

Archeologists have found evidences of tributes to the dead with flowers, in burial grounds dated from the Bronze Age (Despelder and Strickland 2001, p. 42). In Neanderthal burial sites of, 150,000 years ago, ornaments made of shells, stone tools, and food have been found buried with the dead, suggesting the belief that those items would be useful in the passage from the land of the living to the land of the dead. In many of those burial grounds, the body is painted with red ochre and placed in a fetal position, suggesting ideas of revitalization of the body and rebirth (Despelder and Strickland 2001, p. 42).

Death was a central question in Egyptian culture. Pyramids, tombs, mummies, mortuary objects, funerary writings, and the *Book of the Dead*, all provide testimony of a fundamental optimism with respect to death. The *Book of the Dead*, which resembled its Tibetan equivalent, defined funerary practices designed to teach a relatively integrated approach to death (Kastenbaum and Aisenberg 1983, p. 152). This preoccupation with death was reflected in art, religion, and in the sciences of that culture.

Although more primitive societies considered death a natural phenomenon to be accepted without apprehension or fear, this change of focus, with its emphasis on an after-life, and the probable penalties that the dead should suffer after judgment have been passed on them, made it something to be feared. One's own death became psychologically associated with fear of punishment and rejection, the death of others to the fear of retaliation and/or loss of relationship. In more modern times other kinds of fear arise, such as fear of extinction or destruction of the ego (mind) Kastenbaum and Aisenberg (1983, p. 42).

Greek attitudes in relation to death lead us inevitably to the philosopher Socrates. His brilliant student Plato (427–347 AC) gives us, in the *Phaedon or on the Immortality of the Soul* (2004), Socrates' last words, as well as his conversations on death and the dying. Socrates is considered the prince of philosophers. He taught that the purpose of philosophy was to discover the meaning of life in relation to death and to understand the nature of the soul, and that the real philosopher was he who practiced the art of dying all the time. The art of dying, according to Socrates' arguments, was nothing more than accepting death as the separation of the soul (which would continue to exist) from the body (which ceases existing).

For Socrates, fear of death was due to the fact that nobody knew exactly what happened at the moment it occurred. Thus, if the person had no doubts about what really happened at the moment of death, this fear would be groundless. This is confirmed by the many patients who, after having had a near death experience, say they no longer have any fear of death (Noyes 1980; Sabom 1982; Greyson 2007).

> Without the conviction that I will find myself first together with other gods, sages, and the good, and after that, with dead men who are worth more than those of the here and now, I would commit a great mistake by not myself feeling repelled by death (Socrates in Platão 2004).

Socrates believed there was nothing tragic about death and that people should die with an attitude of reverence, thankfulness and peace, with patience and acceptance. But the naturalness of death and the idea of immortality began to suffer the influence of religion, which impregnated the ideas of the popular imagination with suggestions of postmortem punishment and suffering, and even the loss of immortality. The tale of Adam and Eve's transgression of Divine Laws in Paradise and their subsequent punishment is the origin of death seen as a form of punishment, which persists in the religious traditions of Judaism, Christianity, and Islam.

9.3 Death and the Dying in the Middle Ages

Starting at the beginning of the Middle Ages, around the year 400, and continuing for more than 1,000 years thereafter, people living in Western European cultures have shared a vision of the universe as a bond between natural and divine law. The teachings of the Church had a considerable influence over the ways in which people died and nurtured hopes for the afterlife. This period has been characterized by historian Phillipe Ariès (2000a, b, p. 40) as "tamed death." In his understanding, tamed death can be characterized in the following way:

> The ancient attitude in which death is, at the same time near, familiar and diminished, desensitized, represent a stark contrast to our own attitude, whereby our fear is such that we don't dare pronounce its name. This is why, when we call this familiar death the tamed death, we are not implying that it was savage in ancient times and that it was tamed later on. We would like to say, on the contrary, that it has become wild today, while in the past it was not. Death in ancient times was tame.

Kastenbaum and Aisenberg (1983, p. 157) give us a better idea about this period, when they write, in *Psychology of Death*:

> The occurrence of death – the moment of last breath – acquired a new and prohibitive meaning. Now, death was considered God's punishment to man. It was not enough disgrace just to know that he had to die. To complete the story, death revealed man's guilt and indignities, to the extent to which it transported him from a terrifying crisis to eternal torment and mortification.

In the nineteenth century, notwithstanding the light that the Enlightenment cast on various themes of daily life, such impressions still persisted. Allan Kardec (2002, p. 25), a French scholar, provides his critical analysis:

> We have to agree that the picture presented by religion, on that subject, is not very seductive nor consoling. On the one hand, we see the contortions of the damned which expiate their passing mistakes through tortures and spiraling flames. For them, centuries succeed centuries, without hope of kindness nor mercy. And what is even more merciless, their repentance is to no avail.

Death was no more something to be merely contemplated or dealt with through the sacred. It became an event which could be manipulated and molded by human beings. With the decline of the religious view and the rise of the scientific model at the end of the eighteenth century and throughout the nineteenth century, we begin to observe the introduction of a new form of dying, which would not only perpetuate but also enhance the fear of death.

9.4 The Beginning of the Medicalization of Death

At the end of the eighteenth century and the beginning of the nineteenth century, Europe saw the first results of the Industrial Revolution, with the rise of a powerful bourgeois class with its new social, economic, and moral values, the development of more efficient hygienic and sanitary measures, and the construction of great hospitals, equipped with new technology developed by medical research. Those advances reverberated dramatically in attitudes toward death in Western society, making it more distant, impersonal, and devoid of sense with each passing day.

In his book entitled *Western Attitudes toward Death: From the Middle Ages to the Present*, Ariès (2000a, b, p. 310) provides us with a panoramic view of this moment:

> ...An absolutely new type of death appears in the twentieth century, in some more indus-trial, urbanized, and technically advanced parts of the Western world.
> Two traits make themselves immediately apparent to the eyes of the less attentive observer: its novelty, evidently, its contrast to all that precedes it, of which it is the reverse image, the negative; society expels death, except that of men of the State... No one in the city is alerted to anything happening... Society does not stop for a moment: the disappear-ance of one person does not affect its continuity. Things go on as if no one had died.

These changes were to have an effect on society as a whole, but primarily in rela-tion to the dying person and his/her environment. The individual loses control and power over his/her death and is obliged to become dependent on the external envi-ronment. A pact that alternates between neglect and silence, develops amongst diverse groups of people (doctors, priests, scientists, and bureaucratic employees). All are accomplices to a lie that, as it expands, pushes death into the realm of the clandestine (Ariès 2000a, b).

Death does not unfold as it did before, gently, in the presence of loved ones at one's death bed and with the naturalness that it should have. Instead it takes on a savage character, according to the perspective of the French historian Phillipe Ariès (2000a, b, p. 322), to whom we return again for his description of the medicalization of death:

> The room of the dying person went from home to hospital. Due to technical medical causes, this transfer was accepted by families, extended and facilitated by their complicity. The hospital from then on becomes the only place where death can escape publicity – or that which is left of it – from here on in considered a morbid inconvenience. This is why it becomes the place for solitary death.

Thus, the hospital is no longer just the place where people are cured, or die because of a therapeutic failure, but also the scene for natural death, foreseen, and accepted by medical personnel. As a consequence of this medicalization of death, the dying person no longer even has a time to die: The physician is unable to sup-press death, but can prolong life.

Thus, without actually acknowledging it, the hospital begins to offer families an asylum where they can off-load the burden of care for the terminally ill, and con-tinue their own life as usual. Death now neither belongs to the dying, nor to the family, but is regulated and organized by a bureaucracy who assume the responsibil-ity for death and regard it as something that should cause as little disturbance as possible. Death is no longer a natural and necessary phenomenon. It is a failure, it is lost business. Death is now definitively associated not only with fear but also with all that is bad. It is the antithesis of all societal values. Ariès (2000a, b, p. 322), again, expresses this clearly:

> Death no longer instills fear just because of its absolute negativity, but it upsets the heart, just as any other nauseating spectacle. It becomes inconvenient, as do man's biological

needs, bodily secretions. It is indecent to make it public. It is no longer bearable to enter a room that smells of urine, sweat, and gangrene, a room where there are dirty sheets. Access must be prohibited, except for a few intimates who are capable of overcoming its repugnance and to those indispensable care givers. A new image of death is on the path to formation: ugly death, hidden and hidden because it is ugly and dirty.

Apart from forensic medicine, death per se has received little medical attention but has been side-lined as a philosophical matter irrelevant to science. Recent research has tried to restore the balance, introducing it into the medical curriculum, from which it had disappeared since the late nineteenth century (Santos and Incontri 2008; Santos 2009a, b, 2010a, b, c; Feifel 1959, 1977; Kubler-Ross 2001).

It is the dignity of death that is at stake. This dignity demands, in the first place, recognition, not only as a real state but also as an essential event, and one that cannot be belittled.

9.5 Scientific, Psychological, and Philosophical Attitudes in the Face of Death

At the end of the nineteenth century and the beginning of the twentieth, there were a series of attempts to clarify the phenomenon of death. Yet in attempting to understand it, many of the theories suggested only made it more enigmatic than before, deepening its denial and the feelings of fear that surround it. In this short essay, we will analyze the theories that we believe have had the greatest influence over contemporary thought on death.

The theories of Sigmund Freud, considered the father of psychoanalysis, on death and dying reflected the difficulties that he was having in sustaining the fundamental assertion of psychoanalysis that man is an animal oriented only toward pleasure. He created a new theory: the theory of the "death instinct," which maintained that humans had an internal impulse toward death (Thanatos) just as toward life (Eros). This enabled him to explain violent human aggression, hate, and evil in a new way, although still biologically: human aggression comes from the fusion of life and death instincts. Freud's new idea on the "death instinct" was an artifice that enabled him to maintain his earlier theory of human instincts intact, attributing an organic substrate to human evil that ran deeper than the simple conflict between ego and sex. This "new" instinct represents the organism's wish to die, but the organism can save itself from its own death impulse by directing the latter away from itself. For the death wish, is substituted the desire to kill, and man defeats his own death instinct by killing others. Freud would not need to say that death was repressed if the organism took it naturally as one of its processes. In this way, we see that Freud rids himself of the "problem of death" by transforming the latter into a pure "death instinct" (Becker 2005; Freud 1997).

Throughout the nineteenth, twentieth, and twenty-first centuries, three philosophical currents have influenced our way of seeing and dealing with death, and dictated the academic rules with regard to the theme: these are Positivism, Nihilism, and Existentialism.

> Positivism, inaugurated by Auguste Comte, introduces a position in which all who are involved in science limit themselves exclusively to experience... Positive is that which is real, which can be proven through use of the microscope or telescope, that is, scientifically. Science becomes a magic word: it is the new myth that survives today in the cult of material, visible and tangible materials, showing disinterest and even contempt for invisible, intangible values (Comte 1990; Bussola 2000).

Positivism, which descended with a vengeance upon the theology that had dominated the scene and dictated the rules around dealing with death and dying throughout most of human history, fell into the trap of an opposite extremism, which pinned the concept of death to the exclusively material realm or posited it as a biological phenomenon. There are no possibilities for survival after death or transcendence. Not even the introduction of a religion (the positivist religion is the cult of Humanism) was able to spiritualize or lift the human being up. This perspective takes from humans their essence: the Spirit, or in other words, the possibility of spiritual life. Thus, life loses its meaning and being is emptied, giving way to hedonism and the denial of death.

Positivism is the philosophical current that has had the most influence on the sciences, particularly in the field of biology and medicine. In the positivist vision, the human being is a product of chance, constituted merely as an aggregate of atoms and molecules that obey an organizing code: DNA. Thus, death escapes its purview: it is neither analyzed nor taken as an object of study, except as a mere phenomenon that does not raise any particular philosophical issues.

Yet today, with the evolution of medicine, we are witnessing experiences such as those of near-death, which are posing a challenge to the mechanistic model, presenting anomalies that cannot be explained by the prevailing paradigm (Fenwick and Fenwick 2008; Kelly et al 2007, Stevenson 1997; Greyson 2007). Nietzsche, nihilist philosopher and author of the famous phrase, *God is Dead,* engages in harsh criticism not only of religious systems but also of positivist science. In Nietzche, we encounter scorn for the so-called preachers of death, to whom he refers as those with tubercular souls.

> To you I direct praise for my death, a voluntary death that takes me because I want it so (Nietzsche in Morin 1988).

He introduces scepticism and relativism, in place of moral values, ethics, and the ability to reach the truth. In Nihilism, we find death corroding the very concept of itself, as well as all other concepts, undermining support of the intellect, overturning truths, nullifying consciousness. It corrodes its own life, since in a world where all is relative, even the concept of life becomes relative.

Nonetheless, we believe that the philosophical current that has the most influence over the way academics and educators conceive death is Existentialism. This philosophical current tries, notably, to keep anguish alive, to seek within it truths on life and on death. Thus anguish, and consequently death itself, become the best basis for individuality.

For all intents and purposes, anguish can be seen as the common denominator of the philosophies of Kierkegaard, Heidegger, and Sartre. Kierkegaard turns it toward salvation, Sartre, toward freedom, and Heidegger harnesses it to death. Existentialism's major representatives are Martin Heidegger and Jean Paul Sartre. Death is thought of as the end of being's duration over the uninterrupted flux of time. Thus, death becomes the destruction of being, its annihilation.

Within existentialism, death represents the ultimate experience, the one which will provide closure for the individual; for existentialism being is completed through death, after which there is only nothingness. "The authentic Being for death, that is, the finitude of temporality, is the hidden basis of man's historicity" (Heidegger 2000).

Heidegger seeks to eliminate everything that is founded outside of death while Sartre seeks to eliminate all that bases itself on death. Sartre relieves death of its Heideggerian attributes. He snatches from it its irreplaceable nature and its monopoly over the idea of finitude. "Thus it is never death that gives meaning to life; on the contrary, it is what wrests all its meaning from it." And furthermore: "If we must die, then our life has no meaning, since its problems receive no solution and the very meaning of these problems remain indeterminate" (Sartre 2001, p. 35).

In the midst of this general absurdity, "all existence is born without reason, is prolonged out of weakness and perishes by chance" (Sartre 2001, p. 47).

The question that we should now be asking is how to instill dignity in and do away with the fear of death, given the religious, philosophical, and scientific models of the human being that Western Civilization has adopted? To restore this dignity not only to death and dying but also, above and beyond all, to what it means to be a human being, we must discuss philosophical and religious positions, which have now achieved a measure of the scientific support they once lacked. We need, above and beyond all else, the reunification of knowledge, integrating philosophy, science, religion, and scholarship.

Edgar Morin (1988, p. 180) speculates on the clash of nihilistic and spiritualizing currents:

> Yet a truly notable issue, never, in evolved civilizations, has one of these conceptions of death triumphed completely. Neither has persecution been able to destroy, anywhere, the seeds of philosophical religion and of atheism, just as atheism has not been able to destroy the religion of salvation. And this is because each of these conceptions corresponds to a basic human need and because a fundamental contradiction of the individual, between the death that his soul and his being refuse and the immortality that his intelligence repudiates, is unresolved.

Is it possible to think of a hypothesis for its solution? This is what we will discuss next.

9.6   Do We Survive Death? The Final Challenge

If we take a retrospective look at how humanity has dealt with this topic, a number of questions and doubts arise that have apparently remained unanswered throughout history. What can we say about traditional religious understandings and about the narratives of thousands of witnesses of survival of death? Are these and other experiences, such as those of near death, mere fantasies of psychological projections, of self-satisfaction or ego-assertion or the pure and simple result of neurobiochemical reactions? Are these theories able to handle all facts and all anomalies? Does the entire range of concepts of life postmortem have any basis in reality?

If we could come up with a synthesis, as a final analysis, what idea would we work with: Is death a wall or a door? It is essential for humanity to find an answer to this question. Depending upon the answer we find, we may or may not engage in complete modification of our view of the Cosmos and of ourselves, as well as the ethical and scientific implications that this discovery will have in all arenas of human knowledge.

Should we adopt a strictly scientific approach to these issues? Are there phenomena that indicate the possibility of survival after death? Can such phenomena be objects of observation or logical inference?

We believe that science has evolved enough to have a method that, although not infallible, enables us to get close to "reality" and "truth." This method, called the experimental method, should adopt or adapt, naturally, to the object of the analysis that it will take on, that is, the soul. Yet on this type of research, Science has been sorely remiss (must we ask why at this point?) and we are forced to agree with Sommers (1999, p. 5) when he argues:

> If we compare the amount of research conducted on matters relating to life after death to that on any other subject, we must conclude that something is very wrong. It doesn't take a philosophical genius to discover a genuine scandal in the public neglect of the death issue.

Nonetheless, in spite of such omission, a number of researchers over the last 150 years have devoted themselves to this issue. Many of them have produced quality work and results. In our view, an a priori denial or dismissal of this research would be unscientific. Scientists, philosophers, and religious scholars who are interested in this issue should try to study and research these questions from a non-partisan stance. Kardec (2002, p. 22), researcher on these matters, warns:

> ...the knowledge that this teaching brings with it are too deep and expansive to be acquired haphazardly, without serious study and perseverance, in silence and retreat; since it is only in this condition that an infinite number of facts and nuances that the superficial observer will miss, and which enable one to form an opinion, can be perceived.

To qualify as a medical doctor requires 6–8 years of continuous study with support from a number of disciplines, methods and subjects, and then a whole life of practice. So is it not naive to think that just one article or book or one single experiment would be enough to answer the question that has most disturbed humanity: do we or do we not survive biological death?

Before we leave our reader with this challenge, we would like discuss the new studies in this field and theirs implications for the subject of death, mind–brain relationship and survival after death.

9.7   End of Life Experiences: A New Light for an Unsolved Problem?

9.7.1   End of Life Experiences and Their Theoretical and Clinical Implications for Theories of Consciousness

The notion of death as the start of a journey is difficult to sustain since development of Western science, with the concept of consciousness as something that emerges from the brain, and therefore ends with brain death. However, the dying do not always experience the approach to death this way and anecdotal accounts tell a different story often suggestion a continuation of a journey after death. So it is essential to ask the dying what they experience as death approaches. Our study looks at the mental states of the dying and formulates a model of the dying process, which not only has implications for our understanding of death but also more importantly, raises questions about the treatment of the dying, and suggests that compassion is central to their care.

9.7.2   The Research Project

When we (Fenwick and colleagues) started this research project in 2002, very little information was available on the mental states of the dying and very few studies had been published on end of life experiences. (Barrett 1926; Osis and Haraldsson 1986). These end of life experiences we defined as a set of phenomena which occur in the last few days/weeks of life are associated with the process of dying. At the beginning of the study, we were interested in premonitions of dying, deathbed visions, transiting to new realities, terminal lucidity, and the phenomena which occur at and just after death itself – e.g., deathbed coincidences, shapes or light seen surrounding or leaving the body. We were uncertain whether these phenomena truly occurred today, whether they were part of the folklore, or whether, if they did occur, they could be attributed solely to medication or to the organic processes of death itself.

We devised an initial questionnaire to be given to carers of the dying and trialed this with a palliative care team (Fenwick et al. 2009). The results of this study allowed us to construct a questionnaire, which appeared valid and reliable, containing 74 questions. This was then given to two hospices and a nursing home in the south of England, and later to two hospices in the Netherlands and has also been used in an independent study in Ireland by Dr. Una McConville. The study was conducted initially as a retrospective survey, asking the carers to draw on their experiences in the previous 5 years, to get an estimate of the frequency of occurrence, and then as a prospective study. We found that training in this area was very poor and many carers were unaware of the phenomena and thus it became apparent that the initial 5-year retrospective study was also acting as a teaching session for the

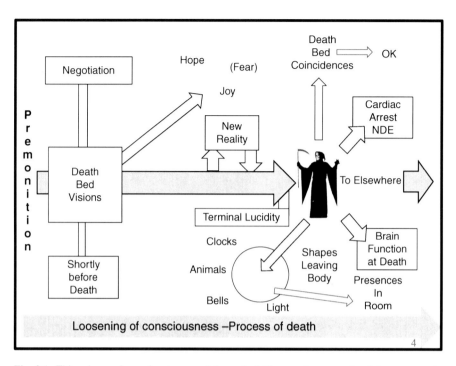

Fig. 9.1 This scheme shows the stages and the end of life experiences as the dying progress to death. Not all people will show every stage. The process may start with premonitions, then death bed visions, transiting to a new reality, terminal lucidity, and then death. At death there may be deathbed coincidences, mechanical malfunction, clocks stopping, animals becoming distressed, light surrounding, and shapes seen leaving the body. Occasionally presences in the room are reported. For completeness, although not discussed in this article, cardiac arrest, NDEs, and post-death brain electrical surges are shown. "Elsewhere" represents what some of the dying mean when they say they are "going on a journey" at death

carers. We returned to the hospice 1 year later and interviewed the same carers a second time in a prospective study, and thus were able to get a much better estimate of the frequency of the end of life experiences, from a population who were now much more aware of the phenomena. As the data were being collected, radio, television, and newspaper articles brought in a total of over 1,000 emails and letters, which gave us a very wide data base, which allowed a qualitative analysis of death-bed visions and coincidences (Brayne et al. 2006, 2008). A summary of our findings is shown in Fig. 9.1.

9.7.3 Premonitions

The Dalai Lama has suggested that premonitions of death are common in eastern cultures, and that people will know they are going to die 1–2 years before their death

(Dalai Lama 2004). Our study is in a Western population and this was certainly not found to be the case. Premonitions were rare, and of three types.

(a) First, and the most unusual, a "knowledge" on the part of the individual that they were going to die. A mother told me this story about her daughter, whose husband was 25. The couple had two young children, but he seemed to be preparing for death. He wrote his will and got his affairs in order, and said that his work was complete and it was now time for him to die. This situation continued for some months and then one day his firm sent him on a trip abroad. The plane that he was flying in crashed and he was killed.

(b) Second and more frequent a premonitory "visit" by – or hallucination of – a dead relative who had come to "collect" them, which may not be recognized as such by the individual.

> We had met at my son's house in August a few years ago because there was to be a small family gathering as it was my granddaughter's birthday. My sister and I were talking (she was 17 years older than I) and she told me that our mother and father had visited her again. They had stood at the bottom of her bed and smiled at her. She was talking to them but they were not answering her just smiling. My mother had died in 1953 and my father in 1998. I was quite upset at the time because I wondered why they had not come to see me! Her son confirmed this because he said he could hear her talking and thought she was dreaming. This had happened to her before. About a month later, she died. Now, if I'm not feeling too well and wake up in the middle of the night, I always check whether anybody is at the bottom of my bed.
>
> My mother was in hospital after a fall and had fractured her ribs. One day she saw my dead brother at the door of her room. She said it had made her very happy to see him. I joked that he had come to take her on to the next world and we laughed about it. Two weeks later she died unexpectedly.

(c) Third, an awareness, perhaps in a dream, that someone close to you is dead, although their death did not in fact occur till later.

> My third daughter was just seven years old when I woke up in the morning totally upset and in mourning over her death. I was so surprised and confused to see her lying next to me that I actually gently nudged her to see if she was alive. She had apparently gotten up during the night and climbed in next to me. When she moved I cried uncontrollably with relief that she was not dead, and hugged her. She did not seem to mind or make any fuss. Perhaps she knew why I felt relief. I did not tell her I had thought she was dead. I could not remember any dream but the feelings were so strong and real, exactly as they were two days later (when) she was the victim of a hit-and-run.

If we accept that true premonitions do occur, we need to postulate that future time leaks backward into the present. There are now a number of experiments that suggest in other situations, and for very short periods of time – 2–3 s – this may be true (Bem 2011).

9.7.4 Deathbed Visions

In Osis and Haraldsson's study of death bed visions, reported by a Western population of US doctors and nurses, religious figures made up 10%, whereas apparitions of the dead were 70%. However, in India, where religion is more central to the culture, religious figures were 50% and apparitions of the dead 29% (Osis and Haraldsson 1986).

In our own study, 60% of carers in the English retrospective and 48% of carers in the English prospective, and 70% of the Dutch retrospective carers reported the occurrence of deathbed visions. Interestingly, approximately 50% of carers in these groups also reported that the apparitions came closer and sat on the bed of the dying. We examined 118 visions and rated a number of parameters. In 24% parents were seen; in 17% persons unknown (usually spiritual); a further 14% who were near to death and unable to speak appeared to greet somebody joyfully. Other relatives and spouses were seen by 14%, angels and friends by only 3%. Importantly, there are occasional reports of the dying saying that they have seen someone they did not know was dead.

The apparitions were seen rarely by others in the room, usually by children, very rarely by hospice or hospital staff. Often they would be seen through the window, waiting, and would usually come into the room shortly before death, stand around or sit on the deathbed. They appear to be located in a particular place as the dying would look at and speak in a particular direction. Often more than one person was seen. The dying reported that the apparitions came to reassure them that they would be with them throughout the death process, and also to tell them when it was their time to go, often setting a date when they would return for the "journey" they were about to take.

The dying respond to their apparitions by showing surprise and pleasure, and after a discussion with the apparition they appear to be ready to leave; the carers reported that after a deathbed visitor had come the dying often used "journeying language," e.g., "When I go," When I am picked up, etc. Much less often the dying person may be afraid or refuse to go, but this is unusual, though sometimes they may negotiate with their visitor to have the time of death delayed for a short while, particularly if a relative was on his way to say goodbye to them.

The obvious explanation for these visions is that these hallucinations are either organic, due to an alteration of brain function by the onset of death, or due to drugs. A number of workers have argued that the deathbed visions are quite unlike the distorted reality of organic hallucinations. For example,

> I looked at the doctors and they seemed quite alright except for their heads. Their heads were different, and as I looked more closely I saw that each had a turkey's head. I knew right them that this was wrong, and it seemed funny afterwards but it didn't go away.
>
> I saw this little man standing on my left. \he was dressed all in grey with a cloak to match. He wore a beard and carried a cane. He just stood and looked at me without making a sound. I knew he wasn't there but he seemed very real.

The hallucinations of TLE are also quite different, although some hallucinations have a very strong feeling of being absolutely real. They are short-lasting – usually seconds – and are usually followed by a spreading of the seizure into other areas of the brain and so to a loss of consciousness. The hallucinations frequently refer to some earlier episode in the patient's life and often to the events surrounding their first temporal lobe seizure. Penfield (1968) stimulated the temporal lobe of patients being operated on for epilepsy and evoked hallucinations.

First stimulation: "yes doctor, yes doctor! Now I hear people laughing, my friends in South Africa....yes they are my two cousins, Bessie and Ann Wheliaw." Another patient stimulated: "I had a dream. I had a book under my arm. I was talking to a man, the man was trying to reassure me not to worry about the book." The stimulus was repeated 1 cm distant she spoke again "Mother is talking to me." Later stimulation at the same point caused her to say. "Yes, another experience, a different experience...A guy coming through the fence at the baseball game. I see the whole thing."

It is clear that brain stimulation evokes only short, transient episodes, which are unlike those of deathbed visions. Another argument put forward is that these are hallucinations due to sleep deprivation. Bliss et al. (1959) reported that after 72 h of sleep deprivation subjects experienced spots before their eyes, undulating tiles on the wall, rolls of luminous chicken wire on the floor, an old lady with gray frizzy hair sticking her head out of the hall, and fine jets of water on the floor. Bexton et al. (1954) reported the mental states of those undergoing sensory deprivations. Their subjects reported that there was movement of simple to complex imagery and that the visual field moved from dark to light. Dots were often seen as were simple geometric patterns and wallpaper patterns. Integrated scenes containing dreamlike distortions also occurred. None of the deathbed visions are thus like these deprivation phenomena.

9.7.5   Drug-Induced Hallucinations

It has been suggested that Ketamine hallucinations could be a good model for near death experiences. This has been looked at in detail by Corraza (2008) in a study of street ketamine users in the UK and Japan. Although there are some similarities to the NDE, there are few similarities to deathbed visions, and so glutamate, an excitatory agonist, which is mimicked by Ketamine and raised during anoxia, is an unlikely candidate as a causative agent for deathbed visions. It has been suggested that in situations of extreme stress DMT (dimethyltryptophan) is released, and might be the causative agent. The experiences of people who had taken DMT have been studied by Strassman (2010). Although all the users had hallucinations, these were not similar to deathbed visions and quite often they related to examination by extraterrestrial monsters. Thus DMT is also an unlikely candidate.

Other drug hallucinations, particularly those following the ingestion of LSD, often contain patterns of bands, spirals, whirlpools or coiled wires, frequently associated with rainbow colors. As the dose of LSD was increased, the colors and shapes became enhanced, then clouds of flower beds, fishtails, long strands of seaweed in a slow moving stream or tiled floors doing tricks were reported; Again unlike deathbed visions.

Mazzareno-Willet (2010) has recently carried out a study on deathbed visions and published a table highlighting the difference between these and other hallucinations.

She emphasizes the response of the dying as being calm, peaceful or elated, the time of the vision, usually months to moments before death, the visions are usually visual, dead relatives and friends, angels, or religious icons. Importantly, she notes that the visions are spiritually transformative and again reports that they are also sometimes witnessed by caregivers, family, or clinicians. Her study adds further support for the view that DBV are in a special category.

9.7.6   Summary

Deathbed visions raise profound questions for models of consciousness. If they cannot be explained easily as solely due to organic processes within the brain, then it would be logical to take the wider view. One such model based on the consistent appearance of dead rather than living relatives suggests, rather than the usual religious explanations of a Heaven or Hell, the possibility of a transcendent realm of nonlocal existence, which comprises simply information. This will be referred to later in the article. Seeing the information pattern of previously living bodies raises interesting questions for a model of consciousness.

9.8   Terminal Lucidity

Terminal lucidity is the sudden arousal into lucid awareness just before death of someone who has been unconscious or semiconscious or demented, so that they are able to greet either people around them or their deathbed visions. There is a long history well reviewed by Nahm and Greyson (2009) and Nahm (2009), which suggests that this phenomenon has been known for many years. We have had accounts from the nursing home of demented patients who had been unable to recognize their family for a number of years, but did so just before they died to say goodbye. It is not clear from a neurophysiological point of view why this arousal should occur as often it is accompanied by behavior, which might not have been possible before. There are accounts of hemiplegic patients sitting up, of patients who had not been able to straighten their backs while sitting and yet are able to do it at this time. The consequence for neuroscience is obvious and clearly this is an area for future research. It would be interesting to know what changes occur in cerebral function during terminal lucidity and whether these changes can account for the phenomenology.

9.8.1   Transiting to a New Reality

In the days before death some patients report that they are perceiving an alternate reality, an area full of love, light, and compassion. Sometimes they are in the hospice, at others in this alternate reality, which appears just as real to them. Few studies have looked at the subjective reports of this phenomenon.

> In the last 2–3 days before she died she was conscious of a dark roof over her head and a bright light. She moved into a waiting place where beings were talking to her, her grandfather among them. They were there to help her. Everything would be ok., it was not a dream. She moved in and out of this area.
>
> My father was at his (my grandfather's) bedside, deeply distressed. My grandfather quietly said to my father 'Don't worry Leslie, I'm all right. I can see and hear the most beautiful things and you must not worry.' He quietly died, lucid to the end.
>
> I was nursing my friend, who had definite views there was no after-life. In her last couple of hours she became very peaceful and arose from her unconsciousness saying periodically such phrases as 'I will know soon,. Come on, get on with it, I am ready to go now,' and 'It is so beautiful.' She would immediately lapse back into unconsciousness after uttering these phrases. She was very obviously content, happy and at peace. It was a wonderful experience for her partner and me.
>
> Suddenly she looked up at the window and seemed to stare intently up at it. This lasted only minutes but it seemed ages. She suddenly turned to me and said 'Please Pauline, don't ever be afraid of dying. I have seen the most beautiful light and I was going towards it. I wanted to go into that light. It was so peaceful. I really had to fight to come back. Next day when it was time for me to go home I said, 'Bye mum. See you tomorrow.' She looked straight at me and said 'I am not worried about tomorrow, and you mustn't be. Promise me.' Sadly she died the next morning…I knew she'd seen something that day which gave her comfort and peace when she only had hours to live.

This new reality contains many of the elements described in the Western NDE – spiritual presences, dead relatives, a bright light, seeing, and hearing beautiful things. It would be logical to suppose that the two phenomena have a common cause. However, in the examples quoted above, the visions were experienced in clear consciousness. There was no evidence for anoxia, hypercapnoea, or other causes to which the NDE is generally attributed. This then raises the possibility that at this very privileged time in the dissolution of consciousness, there may be an unmasking of a transcendental reality and because of the similarity of this reality with that reported by NDEs, it would be logical, applying Occam's razor, to assume they are the same. This then raises questions about the nature of consciousness and the ability to penetrate into one reality from another. This is again in keeping with the possibility of a transcendental reality as postulated by William James. This will be further discussed later.

9.9   Other Phenomena at the Time of Death

A number of other phenomena are reported by carers of the dying at the time or just after death. These are mechanical malfunctions, clocks stopping, deathbed coincidences, shapes seen leaving the body, light surrounding, and transforming the body and abnormal animal responses.

9.9.1   Mechanical Malfunction and Clocks Stopping

Questions about mechanical malfunction and clocks stopping were not asked in our questionnaire. However, a number of accounts have been sent to us in response to media involvement.

'The TV screen went totally blank, the sound went totally and then a nurse came rushing into the room and asked why we had pressed the alarm. At that very moment my father, 58, took his last breath. Nobody had rung the alarm, yet it was ringing in the nurse's office and nobody can explain why the television started malfunctioning. A short while after my father's passing the TV returned to normal. I later spoke to the nurse about the alarm going off and she said it happened all the time at the point of somebody passing.' I am neither a believer or non-believer in an afterlife, but this has certainly opened my eyes to something out there that nobody can understand.

My father died at 3.15 a.m. At about 8.30 a.m. I went to see my Uncle Archie, who'd been close to dad, rather than 'phone him, to tell him about losing dad and bring him back to the house if he wished. As Uncle Archie opened the door it was clear he was distressed and as I began to tell him of dad's passing away he interrupted me and said he already knew…he said no one had telephoned him but told me to look at the clock on the mantlepiece – it was stopped at 3.15, as was indeed his own wristwatch, his bedside clock and all other clocks in the house. There was even an LED display, I think on a radio, flashing 3.15. I was completely taken aback, but Archie seemed comfortable with the phenomena and was just concerned at losing someone close.'

On the morning of my mother's death, I wound her clock, of which she was very fond, and put it to the right time before going to work. While at work a nurse phoned to tell me that my mother had died at ten minutes to eleven that morning. She had not been expected to die that day, and I was very upset that I had not been with her. I drove straight home and the first thing I saw on entering the house was that the clock had stopped at ten minutes to eleven exactly. Twenty-one years later it is still going strong on my mantelpiece.

These examples suggest, if they are correct, that there is a nonlocal field which interferes with mechanical objects at the time of death. These effects are usually limited to the moment of death and thus would fit well with field theories of consciousness, which suggest that there is a marked change in entropy (the dissolution and disorganization of brain and mind and body, discussed later) at the time of death.

9.9.2   Deathbed Coincidences

We have had many accounts of deathbed coincidences and so have been able to examine their characteristics and the mental states of those who perceive them. A recent paper by Haraldsson (2009), studying apparitions in an Icelandic population, reports a number of these to be deathbed coincidences – apparitions of someone close to the individual who at the time they appeared had just died.

In a recent paper, we analyzed 30 deathbed coincidences (Fenwick and Brayne 2011). Seventy-nine percent of the cases occurred at night, which fits in well with the distribution of deaths throughout the 24 h. A third of the respondents said they were asleep at the time the coincidence occurred, and three-quarters were either asleep or in a dream. If the receiver was awake, the experience usually consisted of inexplicable and overwhelming emotion on which the person felt compelled to act. Sometimes a nonspecific feeling was felt. If the coincidence occurred when the receiver was asleep, the phenomena were more complex and the experience tended to be more narrative and specific. A vision of the dying person might be seen or their voice heard. A third received a warning that the sender was dying or had died, and almost half that the sender had come to say farewell and reassure them that they would be alright. Over a third of the recipients described the experience as comforting while a further third found it distressing at the time but comforting later.

In scoring them, we found that not all accounts contained all features, so the total number of cases varied slightly between categories. However, the size of the categories only altered by one or two cases.

Deathbed coincidences raise the question of a linkage, driven by the dying, with a person to whom they are emotionally close. This linkage may extend to the next door room or across continents. Such a linkage is not considered possible by materialistic science as no theory to support communication between minds at a distance is recognized. However, the data that we have make this the most likely explanation as straight coincidence is very unlikely in our data. These experiences could be explained by telepathy or nonlocal linkage, and the work of Dean Radin (2006) and quantum entanglement remains a possibility. It certainly suggests that nonlocal effects do occur.

9.9.3 Light

Light seen surrounding or emanating from the body was reported by a third of the carers in the English sample, although half of the Dutch carers reported this. The light was sometimes described as a transformation of the person and at other times, small globules of light originating from the person would circulate around the room.

> Suddenly I was aware that her father was stood at the foot of her bed. My mother was staring at him too and her face was lit up with joy. It was then that I saw her face appeared to be glowing with a gold light. The light began to leave through the top of her head and go towards the ceiling. Looking back to my mother's face I saw that she was no longer breathing.
>
> Dad passed away in the early 1980's. At the moment of his last breath, I was at his side, holding his hand. I believe my eyes were closed. At this point I can only say that everything disappeared and was replaced by a bright white consuming light. Within this was total peace. No pain, no thought, no time. It could have lasted seconds or minutes. I have no awareness of its duration.
>
> ...odd tiny sparks of bright light emanating from around my brother's body. Not many, just 2 or 3 very brief instances. I did not mention this to anyone else present. However, my brother's wife noticed the same thing and mentioned it, so I told her that I too had seen this.

Light at the time of death suggests either the possibility of bio-photons or of a spiritual light. In favor of a spiritual light is the fact that seeing light has been reported as being viewed with eyes closed, and not by all the people in the room. Spiritual lights are also often described as sparks or balls, which would fit with those seen around the body. Biophotons would be radiated by the collapse of energetic structures disintegrating when the body undergoes anoxic change. This could easily be measured by having photo multipliers in the room with the dying.

9.9.4 Shapes seen Leaving the Body

Sometimes the people near the body say that at the moment of death they saw a shape arising from it. The shapes are never substantial but are described as like mist or smoke or a heat haze. They may come out from the abdomen or crown of the head or, in one account, out of the feet, and will then rise up and disappear.

> ...Gayle came in to tell us Annick had died and we sat around the bed quietly ...what I saw then was totally unexpected. Above Annick's body the air was moving – rather like a heat haze you see on the road but swirling slowly around.
>
> As he died something which is very hard to describe because it was so unexpected and because I had seen nothing like it left up through his body and out of his head. It resembled distinct delicate waves/lines of smoke (smoke is not the right word but I have not got a comparison) and then disappeared. I was the only one to see it. It left me with such a sense of peace and comfort. I don't think that we were particularly close as my sister and I had been sent off to boarding school at an early age. I do not believe in God. But as to an afterlife I now really do not know what to think.

It is interesting that the Dalai Lama has said that in his tradition the dying go through a number of stages before they finally stop breathing. These stages he equates with the loosening and disappearance of different aspects of consciousness. The first stage is that of a mirage, the second smoke, and the third showers of sparks like "the bottom of a wok which sparks when heated over a flame." These three early stages, in his tradition, were the ones most often reported to us. This rather surprising agreement between Western observations of the dying and those of the Dalai Lama's tradition suggest that these features may be transcultural and clearly more research work is required.

9.10 Educational Implications

9.10.1 New Approaches: Death Education-Kinden Garden

Just as death is a taboo in society, the theme is similarly neglected and not taught within schools and universities, especially in health sciences, whether public or private. And if we consider that it is schools that a large part of an individual's cognitive, emotional, psychic, and even existential training is acquired, nothing would be more appropriate than introducing children to the topic early on, preparing them for life which, among other things, means the gains and losses, pains and pleasures, and the acquisition of meaning and purpose in life, and therefore in dying as well (Torres 2008).

Paiva (2008) argues that school, therefore, is the ideal place for the introduction and development of this topic:

> School is the institution that is closest to family. In order for a real partnership to exist between school and family in providing a holistic education for children, schools should create a space to promote information on existential topics, and among them, death, in order to provide orientation for families on how to handle the matter with children. Furthermore, training programs should be offered to prepare teachers to deal with the issue. Schools should take responsibility for education on death.

We understand that the approach to an issue that is as complex as death should be a plural and interdisciplinary one, incorporating the view of a variety of sciences (education, philosophy, theology, anthropology, psychology, medicine, etc.) through interdisciplinary projects in which there is a bond of trust and affection between educator and student, facilitating emotional expression and the learning process, and taking the question of spirituality into account as well, as educators Incontri and Bigheto (2005) assert:

> There are two essential positions on the issue of death. The first, promoted by religions and many spiritualist philosophies, sustains that there is some form of continuity between life before and after death. The second, a materialist one, claims that once physical existence comes to end, so then does being come to its end.

These two visions (materialist and spiritualist) should be given equal space in discussions on death.

> Students' education and the construction of knowledge should be truly guided by a proposal incorporating diverse positions, by the honest debate among ideological currents, the discussion of all convictions that a student should be able to think and know about.

This is emphasized by Incontri and Bigheto (2005) in their article, "Human religiosity, education and death."

9.10.2 Education in Palliative Care

Students in medical schools are taught that medicine is primarily a matter of science and only secondarily about people. Medical doctors are trained to investigate, diagnose, prolong life and cure, yet when a patient is diagnosed with a terminal illness, doctors often feel they have little to offer, beyond feeling profound anguish in the face of the inevitability of their patient's death.

Traditionally, Brazilian medical schools have not allocated time or energy to help their future doctors relate to patients of those who are confronted with death in the short term and their family members. The anguish surrounding death in Western society is reflected in medical education and in doctors' attitudes in the face of death. Quintana et al. (2002), analyzing the problem of death in medical training, after carrying out a series of discussions and arguments, conclude that doctors believe they defend themselves against the anguish over death that their work produces through three mechanisms: negation, rationalization, and the isolation of emotions. Nonetheless, these mechanisms are inefficient and the anguish persists, distorting their communication with the dying patient and family members. Junior et al. (2005), researching medical training and communication with family members regarding death, encountered a difficulty in the first moment of communication, especially when dealing with the death of a young person from an acute condition which the family was not able to understand. Furthermore, they found that 81.1% of professionals saw their academic training for these issues as less than adequate.

In the USA, of 122 medical schools, 100% offer classes on death and dying and 94% offer courses on palliative care. Yet the average number of class hours on death and dying is 12 and on palliative care, 9 h. Considering that medical schools have a total of credit hours of more than 9,000–10,000 h, the amount of time devoted to the topic is practically negligible (Dickinson 2006).

The topic of death is attracting more and more attention on the part of professionals in the health field. Sensitivity around the issue is strongest, at present, among psychologists and nurses, and progress within these professions has become visible, although there is still much to be done, as we will see further ahead. Nonetheless, we believe that Thanatology and the theme of death will only be more fully incorporated within the health field when medical schools place this issue firmly within the undergraduate curriculum and take these issues to professionals working in hospitals and within medicine who are not yet prepared to deal with the finite dimension of life. Doctors as leaders within the health system and as those who are directly responsible for making diagnoses, providing prognoses, communicating critical turns in clinical situations, informing family members when death has occurred and supplying death certificates, have a fundamental role to play in this process of education and consciousness raising. And their participation will have a determining impact in this process. Yet in order for this to become a reality, a Herculean effort is needed, geared not only toward education for death but also above all, promoting reeducation, and thus awakening feelings of humanity, care, compassion, and priesthood in the widest sense of the term, meant to transform medicine from mere technique to art. And in order for this to become a reality we must, in the first place, weed out prejudice, lack of information and ignorance and other obstacles. We can then use the plough of education to begin to plant, the seeds of intellectual, emotional, and spiritual consciousness raising, within the minds of students and particularly among medical professors and directors of medical schools, so that each seedling on the theme of death may be planted and blossom in all medical schools, providing a plethora of fruits.

9.10.3 The Socratic Proposal, the Art of Living, and an Education for Life and Death

Before asking if it possible to educate for death we must ask what education is. Is it possible to teach virtue? To teach the art of life and thus, the art of dying?

Education is an area that has been recognized as interdisciplinary. As Scolnicov (2006) tells us:

> To think about education is to think about complexity, which excludes simple solutions. Theorizing education requires dedication and a willingness to go the difficult road of thought.

In turn, true virtue emerges only as the mature fruit of reflection, which confers an essentially moral character. Not the morals of a particular group or culture, but as a universal ethos that emerges from within each human being and transcends time and space.

Incontri (2007), one of the greatest authorities on Socratic thought in Brazil, has written the following in her chapter, "Being and Death in Sócrates and Plato," part of her book *The Art of Dying – Plural Visions*:

> There is something immanently divine in a being which makes it a moral being. As immortal, the soul has moral conscience. This Ethics is different from the view – common in Greece and even more so today – in which the moral is the sphere of conventional rules that pertain to human groups. Virtue, for Socrates, does not mean obedience to external law, which man concedes, going against his basic instincts. Rather, it is his divine and spiritual nature, the only real source of happiness.

Continuing this line of thought, we find this interpretation of Socratic thought offered by Scolnicov (2006) in his book *Plato and the Problem of Education*:

> This, therefore, is the ethical meaning (beyond its epistemic one) of the superiority of knowledge over opinion. Even if opinion can be made, like an empirical fact, psychologically fortified, there is between it and knowledge a difference which is not pragmatic: opinion, in principle (even if not necessarily so in all cases) is unstable, because it depends on particular conjunctures. Furthermore: whatever stability opinion enjoys will always be the consequence of external forces, fortuitous ones, which have nothing to do with the opinion itself. This is a very Socratic view, that opinion is not **ours**. And which is also to say that popular virtue, unconscious virtue, is not our own.

What Socrates searches for in his Mayeutics is a restructuring and reinterpretation of common values. A life without reflection is a life that is not worth living, even when adorned by all possible practical advantages.

Therefore, to ignore death and its consequences would distance us from the educational proposal that Socrates made and which has unfortunately been ignored in today's society and our universities. What Socrates seeks is not just that we admit our ignorance in the face of the phenomenon of death. He urges us to go further, first destroying our conceptual framework and then reshaping it from the base up. To reflect upon our life means to question our most firmly rooted convictions. It is always a sectarian view that death comes to interrupt or deconstruct. It means to rethink and eliminate contradiction. Death as Socrates interrogated it, through his mayeutics – Does death really mean dying? – poses the Socratic challenge as one for you, here today, my reader!

9.11   Conclusion

Although this is a poorly researched area, a pattern does seem to be emerging. It is not yet clear what is the incidence of the end of life experiences detailed above, what is clear is that they are definitely present and are not excessively rare. When we interviewed the carers, many of the end of life experiences were not discussed in the palliative care team and the hospices, because it was not "safe" to do so. People who spoke about them often felt excluded from the team, as if they were at the "flakey" end of the spectrum. With the more solid evidence now available end of life experiences must be taken seriously. The pattern which we have discussed, shown in Fig. 9.1, does seem to indicate a process. This process does not seem to occur for everybody, and when it does occur may proceed at different speeds and with different emphasis. It is not possible to give materialistic explanations for all the features as some seem to indicate information being available to the dying which comes from remote sources. The recent paper by Bem (2011) highlights the possibility that the brain somehow in some circumstances has access to the future, and would thus point to a possibility that some similar mechanism may be responsible for the occurrence of premonitions. Deathbed coincidences argue for knowledge of distant happenings, which fits in well with studies of telepathy (Radin 2006). In our analysis of 30 accounts, 40% did not know that the person who "visited them" was going to die. The characteristics of the deathbed visions would suggest that they are not simple organic hallucinations in a dying brain, and the need to search for a wider explanation. Here a transcendent reality as postulated by William James could be seen to account for the data. If you ask the dying many in their last hours, believers and skeptics alike, say they are going on a journey and will be collected. All these factors together make a powerful case for consciousness, in some form, beyond the brain. And they certainly suggest that further research into how we die, and how consciousness loosens in the death process, could be a valuable addition to our current models of both death and consciousness.

Acknowledgments   Peter Fenwick: I would like to thank, Sue Brayne, Hilary Lovelace and Ineke Koedam my colleagues who carried out the interviews for the studies.

References

Ariès, P. (2000a). *O Homem perante a morte* (2nd ed., Vol. I). Lisboa: Publicações Europa-América, LDA.
Ariès, P. (2000b). *O Homem perante a morte* (2nd ed., Vol. II). Lisboa: Publicações Europa-América, LDA.

Barrett, W. (1926). *Deathbed visions*. London: Rider.

Becker, E. (2005). *A negação da Morte*. São Paulo: Editora Record.

Bem, D. (2011). Feeling the future: Experimental evidence for anomalous retroactive influence on cognition and affect. *Journal of Personality and Social Psychology, 100*(3), 407–425.

Bexton, W., Heron, W., & Scott, R. H. (1954). Effects of decreased variation in the sensory environment. *Canadian Journal of Psychology, 8*(2), 70–76.

Bliss, E. L., Clark, L. D., & West, C. D. (1959). Studies of sleep deprivation – Relationship to schizophrenia. *A. M. A. Archives of Neurology and Psychiatry, 81*(3), 348–359.

Brayne, S., Lovelace, H., & Fenwick, P. (2006). An understanding of the occurrence of deathbed phenomena and its effects on palliative care physicians. *The American Journal of Hospice & Palliative Care, 23*(1), 17–24.

Brayne, S., Lovelace, H., & Fenwick, P. (2008). End-of-life experiences and the dying process in a Gloucestershire Nursing Home as reported by nurses and care assistants. *The American Journal of Hospice & Palliative Care, 25*, 195.

Bussola, C. (2000). *Introdução ao pensamento filosófico* (7th ed.). São Paulo: Loyola.

Comte, A. (1990). *Discurso sobre o espírito positivo* (p. 4). São Paulo: Martins Fontes.

Corraza, O. (2008). *Exploring the mind-body connection*. London: Routledge.

Dalai Lama. (2004). *Mind of clear light. Advice on dying and living a better life*. New York: Pocket Books.

Despelder, L. A., & Strickland, A. L. (2001). *The last dance – Encountering death and dying* (6th ed.). Moorestown: McGraw-Hill Higher Education.

Dickinson, G. F. (2006). Teaching end-of-life issues in US medical schools: 1975 to 2005. *The American Journal of Hospice & Palliative Care, 23*, 197–204.

Feifel, H. (1959). *New meanings of death*. New York: McGraw-Hill Book Company.

Feifel, H. (1977). *The meaning of death*. New York: McGraw-Hill Book Company.

Fenwick, P., & Brayne, S. (2011). End-of-life experiences: Reaching out for compassion, communication, and connection – Meaning of deathbed visions and coincidences. *The American Journal of Hospice & Palliative Care, 28*(1), 7–15.

Fenwick, P., & Fenwick, P. (2008). *The art of dying*. London: Continuum.

Fenwick, P., Lovelace, H., & Brayne, S. (2009). Comfort for the dying: Five year retrospective and one year prospective studies of end of life experiences. *Archives of Gerontology and Geriatrics, 51*(2), 173–179.

Freud, S. (1997). *Civilisation and its discontents*. São Paulo: Editora Imago.

Greyson, B. (2007). Near-death experience: Clinical implications. *Revista de Psiquiatria Clinica, 34*(Suppl 1), 116–125.

Haraldsson, E. (2009). Alleged encounters with the dead: The importance of violent death. In 337 new cases. *The Journal of Parapsychology, 73*, 91–117.

Heidegger, M. (2000). Being and Time, Harper and Row. (7th ed.). p 220.

Incontri, D. (2007). O Ser e a Morte em Sócrates e Platão. In F. S. Santos (Ed.), *A Arte de Morrer-Visões Plurais* (pp. 71–78). Bragança Paulista: Editora Comenius.

Incontri, D., & Bigheto, A. C. (2005). A religiosidade humana, a educação e a morte. In D. Incontri & A. C. Bigheto (Eds.), *Todos os jeitos de crer* (Vol. 4, pp. 63–76). São Paulo: Editora Ática.

Junior, A. S., Rolim, L. C., & Morrone, L. C. (2005). O Preparo do Médico e a Comunicação com Familiares sobre a Morte. *Revista da Associação Médica Brasileira, 51*(1), 11–16.

Kardec, A. (2002). *O Céu e o Inferno ou a Justiça Divina Segundo o Espiritismo* (10th ed.). São Paulo: Lake Livraria Espírita Allan Kardec.

Kastenbaum, R., & Aisenberg, R. (1983). *Psicologia da Morte* (1st ed.). São Paulo: Editora da Universidade de São Paulo.

Kelly, E. F., Kelly, E. W., Crabtree, A., Gauld, A., Grosso, M., & Greyson, B. (2007). *Irreducible mind: Toward a psychology for the 21st century*. Lanham: Rowman & Littlefield.

Kubler-Ross, E. (2001). *Sobre a Morte e o Morrer*. São Paulo: Martins Fontes.

Mazzareno-Willet, A. (2010). Deathbed phenomena: Its role in peaceful death and terminal restlessness. *The American Journal of Hospice & Palliative Care, 27*(2), 127–133.

Morin, E. (1988). *O Homem e a Morte* (2nd ed.). Lisboa: Publicações Europa-América.

Nahm, M. (2009). Terminal lucidity in people with mental illness and other mental disability: An overview and implications for possible explanatory models. *Journal of Near-Death Studies, 28*(2), 87–106.

Nahm, M., & Greyson, B. (2009). Terminal lucidity in patients with chronic schizophrenia and dementia: A survey of the literature. *The Journal of Nervous and Mental Disease, 197*(12), 942–944.

Noyes, R. (1980). Attitude change following near-death experience. *Psychiatry, 43*, 234–242.

Osis, K., & Haraldsson, E. (1986). *At the hour of death*. New York: Hastings House.

Paiva, L. E. (2008). A Arte de falar da morte: A Literatura infantil como recurso para abordar a morte com crianças e educadores. Tese de Doutorado, Instituto de Psicologia, Universidade de São Paulo, São Paulo.

Penfield, W. (1968). Engrams in the human brain. Mechanisms of memory. *Proceedings of the Royal Society of Medicine, 61*(8), 831–840.

Platão. (2004). *Fédon-Diálogo sobre a alma e morte de Sócrates* (1nd ed.). São Paulo: Martin Claret.

Quintana, A. M., Cecim, P. S., & Henn, C. G. (2002). O Preparo para Lidar com a Morte na Formação do Profissional de Medicina. *Revista Brasileira de Educação Médica, 26*(3), 204–210.

Radin, D. (2006). *Entangled minds: Extrasensory experiences in a quantum reality*. New York: Simon and Schuster (Pocket Books).

Stevenson, I. (1977). Research into the evidence of man's survival after death: a historical and critical survey with a summary of recent developments. *J Nerv Ment Dis, 165*(3), 152–170.

Sabom, M. B. (1982). *Recollections of death: A medical investigation*. New York: Harper and Row.

Santos, F. S. (Ed.). (2009a). *A Arte de Morrer-Visões Plurais* (Vol. 2). Bragança Paulista: Editora Comenius.

Santos, F. S. (Ed.). (2009b). *Cuidados Paliativos-Discutindo a Vida, a Morte e o Morrer*. Rio de Janeiro: Editora Atheneu.

Santos, F. S. (Ed.). (2010a). *A Arte de Cuidar- Saúde, Espiritualidade e Educação*. Bragança Paulista: Editora Comenius.

Santos, F. S. (Ed.). (2010b). *A Arte de Morrer-Visões Plurais* (Vol. 3). Bragança Paulista: Editora Comenius.

Santos, F. S. (Ed.). (2010c). *Cuidados Paliativos-Diretrizes, Humanização e Alívio de Sintomas*. Rio de Janeiro: Editora Atheneu.

Santos, F. S., & Incontri, D. (Eds.). (2008). *A Arte de Morrer-Visões Plurais* (Vol. 1). Bragança Paulista: Editora Comenius.

Sartre, J. P. (2001). *Being and nothingness: An essay in phenomenological ontology*. New York: Kensington.

Scolnicov, S. (2006). *Platão e o Problema Educacional*. São Paulo: Edições Loyola.

Sommers, A. (1999). *Science of human nature, 1*(1), 7.

Stannard, D. E. (1975). *Death in America*. Philadelphia: University of Pennsylvania Press.

Stannard, D. E. (1977). *The puritan way of death: A study in religion, culture, and social change*. New York: Oxford University Press.

Strassman, R. (2010). *DMT: The spirit molecule*. Rochester: Park Street Press.

Torres, W. C. (2008). *A Criança diante da Morte-desafios*. São Paulo: Casa do Psicólogo.

Chapter 10
Research on Mediumship and the Mind–Brain Relationship

Alexander Moreira-Almeida

Abstract Mediumship, an experience widespread throughout human history, can be defined as an experience in which an individual (the so-called medium) purports to be in communication with, or under the control of, the personality of a deceased. Since the nineteenth century, there is a substantial, but neglected tradition of scientific research about mediumship and its implications for the nature of mind. This chapter will review studies investigating the origins, the sources of mediumistic communications. Since one crucial aspect of mediumistic experience is the claim for the persistence of mind activity and the communication of personalities after bodily death, I discuss what would be the evidence for personal identity and its persistence beyond the brain. After that, empirical evidence provided by studies on mediumship is presented and analyzed, including a brief biography of two very productive mediums: Mrs. Leonora Piper and Chico Xavier. Finally, I discuss the implications of these data for our understanding of mind and its relationship with the body. Applying contemporary research methods to mediumistic experiences may provide a badly needed broadening and diversification of the empirical basis needed to advance our understanding of the mind–body problem.

10.1 Introduction

This chapter will review studies investigating the origins or sources of mediumistic communications. First, the definition and the historical and current importance of mediumship to the mind–brain problem will be discussed. Since one crucial aspect of mediumistic experience is the claim for the persistence of mental activity and the

A. Moreira-Almeida (✉)
School of Medicine, Federal University of Juiz de Fora, 36036-330, Juiz de Fora, MG, Brazil
e-mail: alex.ma@ufjf.edu.br

A. Moreira-Almeida and F.S. Santos (eds.), *Exploring Frontiers of the Mind-Brain Relationship*, Mindfulness in Behavioral Health, DOI 10.1007/978-1-4614-0647-1_10, © Springer Science+Business Media, LLC 2012

communication of personalities after bodily death, I discuss some ideas about what would be the evidence for personal identity and its persistence beyond the brain. After that, empirical evidence provided by studies on mediumship is presented and analyzed, including a brief biography of two very productive mediums: Mrs. Leonora Piper (USA, 1857–1950) and Chico Xavier (Brazil, 1910–2002). Finally, I discuss the implications of these data to our understanding of mind and its relationship with the body. The purpose of this chapter is to provide an introductory overview in the form of a short but informative text. More in-depth approaches are provided in many of our references.

There are several possible definitions of mediumship; however, for the purposes of the present chapter, we shall define mediumship as an experience in which an individual (the so-called medium) purports to be in communication with, or under the control of, the personality of a deceased person or other nonmaterial being (Gauld 1982; Hastings 1991; Klimo 1998; Moreira-Almeida et al. 2008; Webster 1996). Frequently, mediumship is regarded as taking place while the medium is in what Bourguignon (1976) refers to as a "possession trance" in which an alleged incorporeal agency takes possession of a medium's volition, speech, and bodily movements. Other common forms of mediumship are hearing, seeing, and automatic writing/psychography (writing attributed to an external, nonmaterial source). A medium's state of consciousness varies along a gradient from totally clear to a full trance in which he/she does not have any memory of what happened during the mediumistic state (Kardec 1986; Braude 2003).

These experiences, through oracles, prophets, and shamans, have been widespread in most societies throughout history, being part of the Greek, Roman, and Judeo-Christian roots of Western society, as well as of Tibetan Buddhism and Hinduism (Hastings 1991). Institutionalized forms of possession and similar forms of ritually induced altered state of consciousness have been described in 53% of 488 societies around the world (Bourguignon 1973, 1976). Currently, there are several social groups that encourage mediumship, such as Catholic Charismatics, Pentecostals, Spiritists, Spiritualists, and many shamans. "Channeling" is a contemporary form of mediumship common in North America (Hastings 1991; Brown 1997).

Mediumistic and trance experiences are usually rich in dissociative behavior, hallucinations, feelings of being controlled by an external power, depersonalization, personality shifts, and posttrance amnesia. Throughout most of the nineteenth and twentieth centuries, trance and mediumistic experiences were regarded as a major cause or manifestation of severe mental disorders by most of the Western scientific community (Almeida 2007; Moreira-Almeida et al. 2005; Le Maléfan 1999). These phenomena were subjects of numerous scientific studies around the turn of the nineteenth century, which were basically divided in two groups, not mutually exclusive:

- Pathological nature of the mediumistic experience and its negative impact people's and society's mental health
- Implications of the mediumistic experience for the understanding of mind and its relationship with the brain

The investigation of mediumship and the discussion of its implications to mind–brain relationships have involved, for more than one century, numerous high level scientists and scholars such as William James (1986; Murphy and Ballou 1960); Frederic W.H. Myers (1903/2001; Cook 1992); Alfred Russell Wallace (Thuillier 1977; Kottler 1974), Cesare Lombroso (1909), Alexander Aksakof (1890/1994); Allan Kardec (1859/1999; Moreira-Almeida 2008), William Crookes (1874; Ferreira 2004), Camille Flammarion (1900/1979), James H. Hyslop (1905a, b), Johann K. F. Zoellner (Stromberg 1989), Gabriel Delanne (1898), Oliver Lodge (1909; Raia 2007), Pierre Janet (1889), C.G. Jung (1983), Theodore Flournoy (1900), William McDougall (Asprem 2010), J.B. Rhine (1937, 1956), Hans Eysenck (Eysenck and Sargent 1993), and Ian Stevenson (1977, 2007). Investigators have the Nobel laureates Charles Richet, Pierre Curie and Marie Curie, J. J. Thomson, Henri Bergson, and Lord Rayleigh (Gauld 1982; Stevenson 1977). Although not well known nowadays, these explorations have provided many contributions to psychology and psychiatry; they were seminal to the development of several of the current concepts on mind such as dissociation, hysteria, and subconscious mind (Almeida and Lotufo Neto 2004; Alvarado 2002, 2003; Alvarado et al. 2007; Crabtree 1993, 2003; Ellenberger 1970; Kelly et al. 2007).

Although for many decades in the last century these studies related to psychical research and spiritualism were presented as a mystical reaction to scientific rationality and other aspects of Western "modernity" in the nineteenth century, more recent historical studies have described them as an integral part of the process of modernization (Aubrée and Laplantine 1990; Lachapelle 2002; Lamont 2004; Monroe 2008; Owen 2004; Raia 2007; Sharp 2006; Treitel 2004;). These studies aimed to use a rigorous scientific and open minded approach to investigate and understand experiences that have been usually approached in two extreme ways: naïve acceptance or dogmatic rejection. Probably the institution that best represented these efforts was the Society for Psychical Research (SPR), funded by Cambridge scholars in 1882 (Gauld 1968). Most of the scientists listed above were members of the SPR, as pointed by Inglis (1982, p. xii): "few organizations have attracted so distinguished a membership." The scientific rigor of SPR is clearly expressed by William James (1897, p. 29): "(…) were I asked to point to a scientific journal where hard-headedness and never-sleeping suspicion of sources of error might be seen in their full bloom, I think I should have to fall back on the *Proceedings of the Society for Psychical Research*."

The question behind the investigations performed by psychical researchers was if the hypothesis that brain creates human consciousness is adequate to explain the full range of human experiences (Kelly and Arcangel 2011; Kelly et al. 2007). These works, however, are virtually unknown today by most current academic authors, and have rarely been used as contribution to the debate regarding mind–brain relationship. Regarding this debate, the study of mediums that claim to be under the influence of deceased personalities is particularly useful because of their potential implication for mind (in)dependence of the brain functioning.

In the last decades, there has been a renewed interest in the study of spiritual experiences such as mediumship, mainly in their relationship with mental health.

It has become clear that dissociative, hallucinatory, and other anomalous experiences are frequent in the general population, and in around 90% of cases are not related to psychotic disorders (Moreira-Almeida and Cardeña 2011). Some hallucinatory experiences provide veridical information suggestive of some sort of extrasensory perception. Based on this, Stevenson (1983) proposed in a paper at the *American Journal of Psychiatry* a new word (*veridical idiophany*) to supplement "hallucination" to designate this sort of nonpathological and veridical unshared sensory experience. There is also evidence that mediumistic experiences often involve people with good or even higher levels of mental health and social adjustment compared to the general population; such evidence does not corroborate the view that the mediumistic experiences are less severe symptoms in a continuum with dissociative or psychotic disorders (Cardeña et al. 1996; Krippner 2007; Moreira-Almeida et al. 2007, 2008; Negro 1999; Negro et al. 2002). There have also been some studies of the neurophysiology of mediumship (Hageman et al. 2010).

Accordingly, recent mainstream studies have explored the relationship of mediumship with mental health, but, despite some revival in the last decade, there has been less attention to the origins of mediumistic experience and their implications for the mind–brain problem (Kelly 2010). Spiritual and other anomalous experiences often happen in altered states of consciousness and they suggest, at least *prima facie*, some sort of mind's independence of brain functioning. Because of this, rigorous and in-depth studies concerning experiences such as mediumship may help to determine the nature of human consciousness, the mind, and their relationship with the brain (Beischel 2007/2008; Eysenck and Sargent 1993; O'Keeffe and Wiseman 2005). As William James (1886, in Murphy and Ballou 1960) has stated about mediumship: "I am persuaded that a serious study of these trance phenomena is one of the greatest needs of psychology" (p. 51) … "it cries aloud for serious investigation" (p. 239).

As we presented previously, dozens of scientists have investigated the implications of mediumship for our understanding of the nature of mind and its relationship to the body. One of the most interesting aspects of mediumistic experiences is the investigation of claims of mind activity after bodily death, what, obviously, would have major implications to our understanding of mind–brain relationship (Eysenck and Sargent 1993, p. 151–2). The understanding of this highly important implication motivated so many high level scientists to investigate mediumistic experiences. They have produced a large amount of data very relevant to our discussion.

The prevalent view in academic world seems to be some kind of reductionist materialistic view of mind, according to which mind is generated by brain activity and disappears with brain destruction (for a historical and philosophical analysis of this claim, please see Chap. 1 of this book). Following the point of view of one of the most important philosophers of science, Karl Popper (1963), this would be a scientific hypothesis, since it is potentially falsifiable. So, the falsifying question is: is there any evidence of continuity of someone's mind/personality activity after brain disintegration? Put in this way, the question is not a metaphysical one, but an inquiry suitable to be answered based on empirical data.

10.2   What Would Be Evidence for Personal Identity and Its Persistence Beyond the Brain?

In line with philosophers of the empiricist tradition such as John Locke and David Hume, I will not focus on the highly controversial issue of the *substance of the mind*. I will not discuss here if mind is a material or spiritual substance, since the notion of substance is a highly problematic philosophical notion. This ontological issue is, at least currently, undecidable. We will focus on the *thinking being* (called mind, consciousness, personality, or soul; whatever its ontological nature) and the phenomenological evidence of its activity.

This kind of exploration needs to be primarily based on a phenomenological approach looking for manifestations of the "thinking being" we are interested in finding vestiges of existence. David Hume (2000, 2007), discussing the project of founding a new "science of human nature," gave priority to the phenomenological level. According to him, the mental phenomena should form the basis of a naturalistic theory of mind, an idea also defended by Locke (1975). Mainly in the controversial and poorly explored area we are discussing in this book, we agree with Hume regarding ascribing epistemic priority to phenomenological theories over theoretical speculations about the hidden causes of the observed phenomena. Phenomenological theories are closer to an observational level than explanatory theories that involve more theoretical speculations about the mechanism involved in explaining the observed phenomena (Chibeni and Moreira-Almeida 2007).

The search for evidence of personality survival after bodily death brings about the discussion of the concept of personal identity, a very controversial issue. Several authors argue that bodily identity is not a necessary condition of personal identity, proposing different criteria for personal identity based on mental or psychological properties (Perry 2008). According to the modern empiricist school (Berkeley 1975), we can know the existence of other minds (in addition to our own minds) in the world only by indirect evidence provided by the experience of certain sorts of observed phenomena suggestive of the operation of other minds. In this way, we can observe evidence suggesting that there are other minds that, like ours, can, for example, wish, feel, and think. We can judge the identity of a mind by a pattern of qualities such as wishes, thoughts, feelings, and ways of dealing with daily matters.

In daily affairs, we use these criteria to identify a friend when we are not seeing him/her, such as in a phone call, or in an e-mail message when there are doubts about who is actually contacting us. As put by Braude (2003, p. 3–4), "we decide who someone is on the basis of what they say and how they behave – more specifically, their memory claim and continuity of character." So, when we cannot have access to the physical part of some personality (the body), we need to base our judgment on this kind of psychological continuity. Quinton (1962/2008) defined soul (personal identity), the essential constituent of personality, "empirically as a series of mental states connected by continuity of character and memory" (p. 59). According to him:

> What is unique about individual people that is important enough for us to call them by individual proper names? In our general relations with other human beings their bodies are for the most part intrinsically unimportant. We use them as convenient recognition devices enabling us to locate without difficult the persistent character and memory complexes in which we are interested, which we love or like. (...) (Quinton 1962, p. 64)

My purpose in this chapter is to investigate if mediumship provides evidence that there is some kind of nonbodily aspect of personality that persists after bodily death. Of course, as in any science, there is not such a thing as a "crucial experiment," the definitive proof beyond any doubt that could not be explained in any other way (Chalmers 1982; Popper 1963). Most scientific experimental results can be explained in more than one way. It is not reasonable to dismiss studies and evidence because they are not perfect. In any science, the best we usually can do is to accumulate good, but not perfect, evidence in favor of some hypothesis and to test whether this hypothesis is falsifiable and can resist falsification. However, the closer the evidence is to the phenomenological level, the stronger the evidence (Chibeni and Moreira-Almeida 2007). The kind of data discussed in this chapter is very close to the phenomenological level. Following Karl Popper's (1963) approach to science, the question of survival could be put in a different way: Is there evidence that falsifies the hypothesis that consciousness is generated by the brain and disappears with physical death? (Moreira-Almeida 2006; Stevenson 1977).

As much as possible, evidence for (or against) a given hypothesis should be based on a wide range observations, covering several kinds of phenomena and not only replications of the same finding. So, the question that we can pose to ourselves now is: what is the sort of evidence that I would accept to falsify the reductionist view of mind and admit the possibility that mind or personality could survive to bodily death? According to Popper, if I cannot state what are the evidence required, my belief in a reductionist mind would be a dogmatic, a nonscientific one, because it would be unfalsifiable.

Cases of children who claim to remember previous life and death-related experiences (discussed at Chaps. 8, 9, 11) are also other phenomena that deserve being investigated in this search; however, in this chapter we will focus only on mediumistic experiences. What sort of evidence would I expect if someone's mind would be able communicate through the body of someone else (a medium)?

Below, I list some of the vestiges of mental activity reflecting a specific personality that seems reasonable to expect if a personality I met in life does not have anymore the body to help the recognition:

- Memory

  - Being able to remember facts, ideally in a large number, accurate, and covering several topics
  - Identify people the claimed personality was acquainted to when alive

- Skills of the alleged personality

  - Speak or write in a foreign language

– Artistic: poem, prose, painting, playing some musical instrument, etc.
– Handwriting

• Personality traits: temperament, character, personal style

Undoubtedly, a large part, probably the most part, of claimed mediumistic communications is composed by generic nonevidential statements. Given the credulity of the sitters, generic and unspecific statements can be easily accepted as evidence of *postmortem* survival by engraved and psychologically fragile people who lost loved ones. Although this is the case for many alleged mediumistic communications, there is a large number that cannot be explained away so easily (Gauld 1982; Almeder 1992; Kelly 2010).

Bellow I summarize the four major hypotheses that have been proposed to explain mediumistic experiences that provide accurate information or other evidence related to some deceased person (Braude 2003; Gauld 1982; James 1909a, b; Moreira-Almeida 2008):

• Fraud, sensory leakage (cold reading, fishing information), lucky chance hits
• Dissociative personality generated by medium's unconscious mind activity involving the access of information stored consciously or unconsciously in medium's memory and the release of latent skills. The emergence of hidden (not ordinarily accessed) memories that might come to surface in altered states of consciousness is called cryptomnesia
• Extra-sensory perception (ESP): mediums deliver information obtained by telepathy from other people's minds and by clairvoyance of distant material sources
• Discarnate mind

The first two hypotheses are compatible with reductive views of mind, but the other two pose serious problems to these views. Of course, these last two options should only be considered after the first two have been excluded as reasonable explanations of a certain observed phenomenon.

10.3 Empirical Evidence Provided by Studies on Mediumship

The large majority of scientists who investigated mediumship in depth ended convinced that conventional explanations (fraud and unconscious mind activity) could explain a lot but not all the observed data. They usually turned to accept the existence of ESP and/or the survival hypothesis (mind can survive bodily death and communicate through another person, the medium) (Eysenck and Sargent 1993; James 1909a; Braude 2003; Gauld 1982; Almeder 1992; Kelly 2010; Bem 2005; Beischel 2007/2008). Of course, there are researchers that remain skeptical on the need of nonconventional explanations to mediumship (Lester 2005; O'Keeffe and Wiseman 2005; see also Moreira-Almeida 2006).

In many stances the medium can provide veridical information known to the deceased personality, but unknown to the medium. These pieces of information may involve details about the circumstances of the death, family nicknames, or incidents known only by one or a few close relatives of the alleged deceased personality communicating[1] through the medium (Almeder 1992; Stevenson 1977). Robertson and Roy (2001) conducted a study to answer the question: is it true that the statements made by mediums to recipients are so general that they could be accepted just as readily by anyone? They studied ten mediums during 2 years and took notes of the specific statements made by mediums to 44 sitters (recipients of messages) unknown to the mediums. These 44 recipients were asked to identify statements relevant to them in the set of statements specifically made to them. The 44 sets of statements were also submitted to 407 nonrecipients who also rated the relevance of the statements to them. The median percentage of acceptance of the statements among nonrecipients was 30%, while among the recipients the average was 65%, which was highly statistically significant ($p = 5.37 \times 10^{-11}$). So this study does not support the hypothesis that mediums' statements are so general that would be accepted by anyone. However, one important limitation of this study is that it was not double-blinded, which could have allowed mediums to deduce information and influence recipients in their ratings. Because of this, double-blind protocols (mediums do not have direct contact with sitters, and sitters rate control and direct communications to them without knowing which is intended to have been proposed (Roy and Robertson 2001) and tested with some positive (Roy and Robertson 2004; Schwartz et al 2002) and negative results (Jensen and Cardeña 2009; O'Keeffe and Wiseman 2005; Schwartz et al. 2003)). However, several methodological limitations to these studies have been pointed (Kelly and Arcangel 2011). One of the major challenges facing this kind of research (as well as other studies in psychology and psychiatry) is to balance between rigor and providing a more naturalistic environment for the study that allows the target phenomena to occur. Specifically regarding mediumship, it is necessary to avoid sensory leakages that could be clues for the medium, but also it is necessary to take in account the mediums' need of an environment that allows them to work in a comfortable way, to be able to connect with the communicating personality (Kelly and Arcangel 2011; Jensen and Cardeña 2009; Beischel 2007/2008). Also, based on previous studies, it is necessary to take in account that mediums do not get similar levels of veridical information compared with each other and even the same medium in different occasions (like any other human skill). It is also important to focus studies not on a random sample of mediums, but on more gifted mediums, those who have already consistently provided reasonable evidence suggestive of anomalous information reception (Beischel 2007/2008). It is in line with William James' (1886) suggestion to focus mediumship research on "good specimen of the class" (p. 97).

One research strategy that has been employed is the "proxy sitting," where "the person desiring a communication from a deceased loved one is not physically present at the sitting; instead a third person, preferably someone with little or no

---

[1] To avoid the burdensome and long expressions such as "alleged communicating personality" or "claimed communicating personality" and to make the text easier to read, I will use just "communicating personality" or some similar expressions, not necessarily meaning that I agree that this personality actually is what it claims to be.

knowledge about the deceased person, arranges the sitting with the medium and attends it as a proxy for the real sitter" (Kelly 2010, p. 256). It has the advantage of eliminating the sitter as a source of information (by cold reading or telepathy) and allows a sitter to blindly evaluate and compare the accuracy of the reading intended to him/her with control readings.

I will now discuss the two most rigorous recently published researches evaluating anomalous communication in mediumship using proxy sitters. Beischel and Schwartz (2007) performed a "triple blind study," in which "(a) the research mediums were blind to the identities of the sitters and their deceased, (b) the experimenter/proxy sitter interacting with the mediums was blind to the identities of the sitters and their deceased, and (c) the sitters rating the transcripts were blind to the origin of the readings (intended for the sitter vs. a matched control) during scoring" (p. 24). The research question was: "Can research mediums obtain and report accurate and specific information about targeted deceased individuals (discarnates) when both the mediums and the experimenter/proxy sitter are blind to information about the sitter and discarnate during the reading and the raters are blind to the origin of the transcripts during scoring?" (p. 24). Eight mediums provided information about specific deceased relatives of eight students of the University of Arizona. Mediums were approached by proxy sitters who knew only the first name of the deceased personality. Each medium was asked to contact two deceased personalities. Each blinded sitter rated (from 0 (No correct information or communication) to 6 (Excellent reading, including strong aspects of communication, and with essentially no incorrect information)) the specific statements made in two readings (one intended for him/her and a gender-matched control reading) and chose the reading more applicable to him/her. The average score for intended reading was about twice (mean 3.56) the control readings' scores (mean 1.94) ($p = 0.007$). Three mediums got a very high average score for their readings (between 5 and 5.5), two obtained moderate average rates (3.5), and none of the mediums produced readings that were scored lower than control ones. Sitters identified the correct reading 81% of the time (13 of 16, $p = 0.01$). The study controlled for potential sources of fraud, sensory cues, and rater bias. Authors concluded that "under stringent triple-blind conditions (…) evidence for anomalous information reception can be obtained" (p. 26).

Kelly and Arcangel (2011) recently published in a traditional psychiatric journal two studies testing different methods. The first and smaller study did not produce statistically significant results, which authors attributed to methodological problems that would not allow mediums to focus on the deceased person and create difficulties for sitters in rating the readings. Based on these findings, they performed the second and larger study with nine mediums and 40 sitters, using two proxy-sitters. Mediums received only a "neutral" photography of the deceased ("showed the person alone and not engaged in specific activities that might provide significant information about the deceased" p. 13), and provided the reading by phone to a proxy sitter. All readings were transcribed. Each sitter rated blindly six reading transcriptions (one reading directed to her/his deceased relative and five other directed to other sitters' relatives, but matched by communicator's gender and age). Fourteen of the 38 returned sets of ratings were correctly chosen (ranked as first among the six readings) and seven were ranked as the second. Thirty of the 38 readings were ranked in the top half ($p < 0.0001$). As in other studies, some mediums performed much better than others. For one medium, all of her six readings were ranked as number 1. In evaluating mediumistic communications, it is important to take in consideration not only quantitative results, but also the qualitative nature of the data obtained to better evaluate the value of them. For example, one of the 14 people who correctly chose their own reading said: "I feel certain this is the correct choice and would bet my life on it" (p. 14). Among the reasons this sitter provided for his choice was "the medium's statement that 'there's something funny about black licorice.... Like there's a big joke about it, like, ooh, you like that?' According to the sitter, his deceased son and his wife had joked about licorice frequently. Also, the medium had said 'I also have sharp pain in the rear back of the left side of my head in the back, in the occipital. So perhaps there was an injury back there, or [he] hit something or something hit him.' The deceased person had died of such an injury incurred in a car crash." (p. 14). It is quite remarkable that even under these controlled and artificial conditions mediums on average (and specially some of them) were able to provide several specific pieces of information that were recognized by blinded sitters. Authors pointed that "this study might suggest to readers that mediums are neither the infallible oracles that many people in the general public seem to believe they are, nor the frauds or imposters that many scientists assume they invariably are" (p. 16).

Using the methodological principle of Ockham's razor ("entities are not to be multiplied beyond necessity"), several researchers claim that veridical information provided by mediums could be explained by ESP, not needing to require the existence of a discarnate personality. According to this position, mediums could, for example, obtain this information telepathically (often unconsciously) from the relatives who seek them looking for a mediumistic communication. Although certainly applicable to many situations, it becomes more improbable when the communicating personality provides veridical information unknown by the sitters or even when there is no sitter acquainted to the alleged communicating personality. There are cases in which the communicating personality is unknown to medium and sitters at the time of the communication, appears spontaneously and provides specific details that are later verified as accurate. This is what was termed "drop-in communications" and is considered by some authors as providing strong evidence for the survival hypothesis, since it would be harder explaining these kinds of events in terms of telepathy or clairvoyance. This would be specially the case when "the verified information contained in the communication was not obtainable through any single source" (Kelly 2010, p. 255). Another relevant aspect of drop-in cases is the *motivation* for the communication. In these cases, it is not clear what would be the motivation of medium and sitters to obtain by extrasensory perception information of someone unknown to them. On the other side, in these cases, the communicating personality usually has a purpose to communicate, often console a relative in mourning or take care of some unfinished business (Braude 2003). There are some drop-in cases reported in the literature that deserves consideration (Gauld 1982; Haraldsson and Stevenson 1975a, b; Kardec 1859a, b, 1863; Ravaldini et al. 1990; Stevenson 1970, 1973).

The philosopher C.J. Ducasse (1962) was one of the first to argue that it is even harder to explain by ESP the cases where the communicating personality exhibits not only pieces of information (knowledge *that*) but also skills (knowledge *how to*) unlearned previously by the medium but possessed by the communicating personality when in life. The reason is that skills could be acquired only by practice. It is not enough to learn about music and piano to able to actually play a piano. In this sense, a rare and challenging phenomenon is xenoglossy, when mediums speak in real languages that they do not know. Well documented cases of xenoglossy are rare and although the communicating personality may be able to talk in another language he/she usually is not completely fluent in the new idiom (Almeder 1992; Bozzano 1980; Gauld 1982; Stevenson 1974, 1976, 1977).

Other kinds of unlearned skills displayed occasionally by mediums are xenography (writing in a real but unlearned language), painting, drawing, and poetry (by mediums who had no prior training and do not exhibit those skills in their ordinary lives). Finally, some mediums channel by writing with a handwriting similar to the alleged communicating personality when that person was alive, a still poorly studied phenomena (Gauld 1982; Kardec 1858a, b, c, Kardec 1860, 1861/1986; Perandréa 1991).

Some mediums also showed a wide range of psychological traits (character, humor, conciseness, mannerisms, choosing of words, likes and dislikes, etc.) related to the communicating personality, providing a more convincing impersonation of the communicator (Braude 2003, Hodgson 1898). This is what Hyslop (1905a) termed "selective unity of consciousness," and what he considered a "positive argument for the spiritist hypothesis" (p. 268). It was also proposed by Ducasse (1962, p. 405) as "particular modes of active thinking which *only* the particular mind whose survival is in question is *known* to have been equipped with." The emphasis here is on not only factual information (knowledge that) provided by mediums as evidence for survival of consciousness after death, but also the demonstration of unlearned skills and a broad variety of personality traits typical of the personality when alive has been emphasized by other researchers such as Ian Stevenson (1977) and Robert Almeder (1992). Stevenson et al. (1989) published an interesting case of possession in which the secondary personality provided not only evidence of knowledge, but also a complex set of behaviors and skills characteristic of Shiva, a woman unknown to the medium and her relatives who lived 100 km away and who had been murdered 2 months earlier. This possessing personality was able to recognize 23 persons known to Shiva, displayed knowledge, and a wide range of behavior accorded with Shiva's personality such as style of dress, caste snobbery, increased literacy, and tendency to joke.

However, as stated previously, most of mediumistic communications do not provide some kind of accurate and verifiable phenomena that deserves being investigated as evidence of survival or even ESP. Actually, even the best mediums alternate between "good" and "bad" days. Usually, even a good medium in a good day often produces a mixture of veridical information with irrelevant data, just as eye-witness observers of current life situations describe events with at best partial accuracy. Nevertheless, good mediums consistently produce a high amount of accurate and relevant phenomena (Gauld 1982; James 1909a; Eysenck and Sargent 1993).

If we consider the hypothesis that mediums can actually display ESP and/or have the capacity to transmit information received from discarnate minds (what would also be a sort of telepathy), it is reasonable to suppose that this skill, like most of other human capacities (playing soccer, being a humorist, composing a poem or song, etc.) is dependant of several internal and external circumstances (Hodgson 1898; Braude 2003). Sometimes, because of neurological, emotional, pharmacological or even environmental conditions, I (my mind) cannot express myself in an easy, clear, and unimpeded way. If we consider the hypothesis that mind might in some way survive bodily death, it could have even more troubles in communicating through medium's brain and body. One analogy would be trying to talk to someone else using a bad and noisy mobile connection with crosstalk. Several authors proposed that mediums may receive this anomalous information in a discontinuous way, fragmentary pieces of information, many times through mental imagery with symbolic meaning (Alvarado 2010). This kind of consideration and the available data led several researchers to conclude that even in veridical mediumistic communication, the gaps between information obtained from a discarnate mind would be filled by contents obtained from the medium's mind and ESP. According to them, the mediumistic dissociative state would facilitate the emergence of subliminal mind content, but would also allow ESP and contact with a discarnate mind. This hypothesis was developed more fully by Frederic Myers (1903/2001), but was also accepted several others (Braude 2003; Almeder 1992; James 1909a; Gauld 1982; Kelly and Arcangel 2011; Alvarado 2010).

Other kinds of mediumistic phenomena that have been investigated are crosscorrespondence (different mediums, with no ordinary contact between them, produce mediumistic communications that, by themselves, do not make sense, but when put together make sense), death bed visions and other apparitions by the time of death (Gauld 1982; Stevenson 1977; Kelly 2010), and physical manifestations such as materializations and movement of objects. These last, the physical phenomena, were subject of many instances of fraud, what raised strong suspicion regarding this kind of manifestation (Gauld 1968, 1982).

10.4 Brief Report of Two Mediums

Data discussed until now in this chapter deal with evidence provided by mediums; however, a general idea about the biography of some mediums can also help to better understand mediumship and its implication for the mind–brain problem. To provide some details about particular mediums and their mediumistic skills, we will pass to a brief description and the analysis of two prolific mediums with different characteristics: an older, very well documented, and investigated case (Mrs. Leonora Piper 1859–1950) and a contemporary but much less studied (Chico Xavier 1910–2002).

10.4.1 Mrs. Leonora Piper (USA, 1857–1950)

Mrs. Leonora Piper is probably the most studied medium and one of those who produced more evidence suggestive of an actual communication of a deceased personality. Literally, thousands of pages were published with reports of her séances and analysis performed by a wide range of high level scientists (Hodgson 1892, 1898, Lodge 1909; Hyslop 1905a; Sidgwick 1915)

She lived in Boston and started having visions when she was 8 years old. In 1884, attending a mediumistic meeting as a sitter with another medium, she unexpectedly fell in trance and wrote a letter to another sitter (the Judge Frost) attributed to his deceased son. After then, she started to act as a medium, first providing communications by voice and later mainly by writing (Hyslop 1905a).

She had what was called "controls," personalities who controlled and supervised her mediumship. She had four major controls, each of them during a given period: Phinuit (1885–1892), G.P. (George Pelham) (1892–1897), Imperator (1897–1905), and Richard Hodgson (1905–1911). Phinuit and Imperator were very probably artificial personalities created by her mind, on the other side, G.P. and Hodgson provided much more and stronger evidence of being an actual personality. As usual with other mediums, controls often acted as the spokesman of communicating personalities, often being the intermediary between the sitter and the communicating personality (Gauld 1982; Hyslop 1905a; Sidgwick 1915). As usual even in the most skilled mediums, there was a high fluctuation in the accuracy of the information she provided. Usually the communications were a mix of precise and imprecise information (Hodgson 1898).

The Harvard philosopher, physician, and psychologist William James was the first to systematically investigate Piper's mediumship. In the beginning, to evaluate if her mediumship was worth of further investigation, he took 25 different sitters under pseudonyms to her. During trance, she displayed an impressive knowledge of private affairs of sitters that were unknown to other people present at the meetings. He realized that she was worthy of further investigation, then he and other researchers started an in-depth and continuous investigation. She was able to provide a large number of precise facts that could not be explained away by conventional means (Hodgson 1898, Hyslop 1905a; James 1890, 1896, 1909). After more than 10 years of investigations of Mrs. Piper's mediumship, James stated (1896/1986):

> (…) a universal proposition can be made untrue by a particular instance. If you wish to upset the law that all crows are black, you must not seek to show that no crows are; it is enough if you prove one single crow to be white. *My own white crow is Mrs. Piper.* In the trances of this medium, I cannot resist the conviction that knowledge appears which she has never gained by the ordinary waking use of her eyes and ears and wits. What the source of this knowledge may be I know not (…); but from admitting the fact of such knowledge I can see no escape. So when I turn to the rest of the evidence, ghosts and all, I cannot carry with me the irreversibly negative bias of the rigorously scientific mind, with its presumption as to what the true order of nature ought to be (p. 131) [italics added].

After James' initial studies of Mrs. Piper James, Richard Hodgson, a friend of him, took the lead on the investigation. Hodgson was a skeptical member of the SPR, considered an expert in unmasking fraud; his strength was hostile and aggressive debunking. He moved to Boston in 1887 to investigate Mrs. Piper, he introduced sitters anonymously or pseudonymously, used proxy sitters, made complete word-for-word records of séances, and took signed testimony from the participants. Hodgson even hired detectives to follow Mrs. Piper. SPR investigators also took her to England, where she knew no one and was kept under strict surveillance and had her baggage checked. Even under those conditions, she continued to provide repeatedly accurate information (Hodgson 1892, 1898; Gauld 1982; Hyslop 1905a; Lodge 1909).

In testing the veracity of the communicating personality G.P. (George Pellew), Hodgson introduced 150 people to G.P., but only 30 were actually acquaintances of G.P. when alive. The communicating G.P. was able to recognize 29 of these 30, and only among those 30. He interacted properly with each of these 29 and showed knowledge of facts known only by each sitter and G.P. After years of studies with Mrs. Piper, Hodgson became convinced that "communicators were, at least in many cases, what they claimed to be, namely the surviving spirits of formerly incarnate human beings" (Gauld 1982, p. 34).

In 1905, Hodgson died suddenly and Mrs. Piper started to have communications attributed to him. Although the period when Hodgson was the control was not as evidencial as that of G.P., it also provided a lot of evidence of anomalous information. James (James 1909a, b, p. 209) produced a long report of this period, concluding that:

> I myself feel as if an external will to communicate were probably there, that is, I found myself doubting, in consequence of my whole acquaintance with that sphere of phenomena, that Mrs. Piper's dream-life, even equipped with 'telepathic' powers, accounts for all the results found. But is asked whether the will to communicate be Hodgson's, or be some mere spirit-counterfeit of Hodgson, I remain uncertain and await more facts (...)

Mrs. Piper's is a remarkable case since she was deeply investigated by dozen of scientists for almost 25 years, and there was never found any reasonable evidence for fraud. There are thousands of pages registering her séances. She produced not only abundant and accurate information, but also mannerisms, verbal expression, and sense of humor compatible with the dozens of the alleged communicating personalities. Every investigator who studied her in depth became convinced that conventional explanations were not enough to her and that some kind of anomalous process (usually ESP and/or personality survival) must be involved (Gauld 1982; Eysenck and Sargent 1993; Hyslop 1905a; Lodge 1909; Sidgwick 1915).

10.4.2   Chico Xavier (Brazil, 1910–2002)

Francisco Candido Xavier, known as Chico Xavier, started having visions of deceased people, mainly her late mother, when he was 4 years old. He was raised in a very poor and large Catholic family in the countryside of Brazil. He received only elementary education (until fourth grade), started to work very young (8 years old), and lived in a place with difficult access to libraries.

When he was 17 years old, he started contact with a spiritist center and started receiving communications by automatic writing, also called psychography. In 1932, his first mediumistic book was published, with dozens of poems attributed to several Brazilian and Portuguese dead poets (Souto Maior 2003).

Chico Xavier produced a wide range of mediumistic phenomena; some of the most interesting were books and letters produced by automatic writing. He never accepted any payment or gifts for his mediumship, which he understood to be a spiritual gift that should be used as a tool of charity to help suffering people. The copyrights of all his books were donated to charity organizations. He always kept a regular modest job, and always lived a very modest life. He has been a major impact in Brazilian culture, a movie about him brought to theaters more than 3.4 million people, being the third most popular Brazilian movie in 2010 (SDRJ 2011).

He produced by automatic writing more than 400 books covering a wide range of styles and topics: historic novels, history, physics, biology, ethics, religions, short stories, poems, etc. Many of these books are best sellers translated to more than ten languages. His books sold around 30 million copies; only one of them (*Nosso Lar*) sold more than two million copies and was made into a movie that attracted more than four million spectators in 2010 (Matos 2011; SDRJ 2011). Unfortunately, Chico Xavier and his mediumistic production have been subject to little research. We will discuss some of them and hope that more can be done in the near future.

His first book, *Parnaso de além-túmulo* (Anthology beyond the tomb), is a collection of 60 mediumistic poems attributed to 14 Brazilian and Portuguese deceased poets that was published by the first time in 1932. The following editions incorporated more poems attributed to other poets. The definitive edition (6th) was published in 1955 and contains 259 works attributed to 56 Brazilian and Portuguese poets, 27 of them little known or even unknown. This book was studied in depth by an expert in Brazilian literature (Rocha 2001a, b). He selected three Portuguese poets (João de Deus, Antero de Quental, and Guerra Junqueiro) and two Brazilians (Cruz e Sousa and Augusto dos Anjos) present in the book to analyze the similarities between the works of these poets when alive and the poems attributed to them at the anthology. His main question was: "are the authors' poetic voices convincingly recovered by the poems?" Rocha analyzed the stylistic, formal, and interpretative aspects. The conclusion was that the analysis suggested that the poems of the anthology are not a product of simple literary imitation.

I will describe briefly the analysis (Rocha 2001b) of the six poems attributed to the poet Guerra Junqueiro (1850–1923) that were published between 1932 and 1935, one of them is the continuation of a poem left unfinished by Junqueiro in life. Rocha used as reference the best work on Junqueiro ever written (Carvalho 1945). This work, published at least 10 years after Chico Xavier wrote the poems, identified the most important Junqueiro's characteristics: Symbolization and personification, satire and caricature, use of type-figure and symbol-figures, bucolic and nostalgic feelings, alexandrines, and internal rhymes. Nostalgia is the only characteristic that is not present in the poems psychographed by Xavier. Carvalho also identified six stylistic features that characterize the subjectivity of Junqueiro, all of them are present at the mediumistic poems: tendency for animism (attribution of conscious life to phenomena of nature or to inanimate objects) and personification, intense use of light, color and gold, multi-adjectivation, use of synonyms and repetitions, antithesis and use of opposing ideas, and rhythm. In summary, the author of the psychographed poems was able to skillfully synthesize in only six poems basically all Junqueiro's characteristics and style. It is worthwhile to remember that these poems were written by a young and poorly educated medium living in the country-side of Brazil in the beginning of the twentieth century, which means a very restricted access to education and literacy. The case becomes even more challenging if we remember that Junqueiro was only one of the 56 poets represented at the book Parnaso. In depth analysis of another four poets can be found elsewhere (Rocha 2001a).

Another study was performed on the mediumistic writings attributed to the Brazilian writer Humberto de Campos (1886–1934). He was a prolific author, writing hundreds of journal articles and publishing about 45 books. Less than 4 months after his death, Chico Xavier start to write texts attributed to him. Twelve mediumistic books were published between 1937 and 1969, selling more than one million copies. A Literature PhD dissertation investigated the mediumistic writings attributed to Humberto de Campos (Rocha 2008). The author read the books published by Campos in life and those attributed to him. The conclusion was that the author of the mediumistic books knew in depth Campos' works and was able to reproduce Campos' style and character. The mediumistic writings attributed to Campos are full of both subtle and explicit references to the works he produced when alive; there is an intricate and sophisticated intertextuality that is detectable only by those who know very well Campos' works. The mediumistic Campos makes adequate citations of many authors as well as historical and mythological characters that were part of the cultural repertoire of Humberto de Campos when alive. An interesting aspect is that some of this information makes references to Campos' writings that were not in public domain when the mediumistic texts were produced. For example, they refer to the "Secret Diary," that was kept locked in a safe at the Brazilian Academy of Letters until 1954, 20 years after Campos' death (Rocha 2008).

Another important but less studied aspect of Xavier's mediumistic production are the letters that deceased personalities used to write to relatives and friends left behind. Chico produced thousands of these letters. Most of them were produced in public meetings in a spiritist center where dozens of people from all over the country went hoping for receiving such a letter. Some of these people were acquaintances of Chico, but most of them were not. Usually, before starting to write these letters, he talked to these people in line for a few minutes in a public room. Although there has not been yet a very rigorous research on this letters, preliminary reports point to findings that deserve further investigation. Even if many letters contained just generic statements or pieces of information that may have been passed to Xavier in the short talking he used to have with the relatives before writing the letters, there are also other findings that seem to need other sort of explanation. Often these letters contained accurate personal information regarding the communicating personality, his/her family, lifestyle or death, some of these details unknown by the relatives who had talked to Xavier and received the letter. There are also reports of letters written entirely or partially in a foreign language and of handwriting similar to that of the deceased when alive (Severino 1994). Our research group has started in 2011 an in-depth investigation of these letters.

A study about the handwriting identity of four letters written in Italian by Xavier in 1978 and attributed to Ilda Mascaro Saullo, who died in Rome in 20 Dec 1977 was also published (Perandréa 1991). The researcher, an expert in handwriting identity, compared the handwriting of that psychography with Ilda's handwriting when alive, Xavier's regular handwriting and with other messages produced by Xavier's automatic writing. He concluded by the handwriting identity between the psychographed letters and the writings of Ilda when alive. I am not aware of other studies in this line and of other handwriting experts' opinion regarding the presented study. So, although these conclusions need to be analyzed carefully, this study points to another important line of research regarding mediumship.

10.5 Conclusion

Mediumship is a sort of human experience that has already contributed to our understanding of mind, having a major role in the development of concepts such as dissociation and subconscious mind. Although it has been neglected for several decades, the open-minded, rigorous, and creative scientific investigation of mediumistic experiences is certainly a major tool to move forward our exploration of mind and its relationship with the brain.

In spite of scarcity of funds and lack of institutional support of most academic organizations, research on mediumship have mobilized dozens of distinguished scientists for more than one century and reached to considerable results. The few recent and well controlled studies have replicated the previous findings of past psychical researchers that mediums, even under strict controlling conditions, can obtain some sort of anomalous information regarding discarnate personalities. Mediums in trance have also been able to exhibit skills beyond those displayed in normal states of consciousness. To better understand human nature, we need to explore mediumship in more depth and to develop and test theories to explain these challenging phenomena.

We need to be aware of the risk to assume that experiences based on superficial similarities are identical. This may lead to unwarranted inferences of causes and of physiological substrates. One example of this was taking mediumship as a manifestation of psychosis or dissociative identity disorder (Moreira-Almeida et al. 2005, 2008). Undoubtedly, mediumship is a very complex experience that probably involves ontologically different sorts of phenomena and cannot be fully understood by one single explanatory hypothesis. In depth, comprehensive and interdisciplinary, studies of specially gifted mediums may be very useful in advancing our understanding of mediumistic experiences.

The evidence available regarding mediumship strongly suggests anomalous sources of information; these data are anomalies to the paradigm, potential falsifiers of the reductionist model for the mind–brain problem. It is very hard to account for all the existing data without taking in consideration ESP and/or the survival of the disincarnate mind after bodily death. This was the conclusion reached by most of those scientists who studied in depth mediumistic experiences. I, in line with several authors (Almeder 1992; Braude 2003; Eysenck and Sargent 1993; Gauld 1982; Kelly et al. 2007; Myers (1903/2001); Stevenson 1977), think that the whole body of existing data seems to be better explained by the hypothesis of the persistence of mental or spiritual activity after bodily death. This is in line with William James (1898) "transmission theory" of cerebral action, in which "brain *can* be an organ for limiting and determining to a certain form a consciousness elsewhere produced" (p. 294). This hypothesis becomes even more consistent if we take in consideration several other phenomena discussed in this book, such as near-death experiences and apparent cases of reincarnation.

Anyway, both of those hypotheses (ESP or survival) cannot be accommodated within the view that mind is just a product of brain chemical and electric activities, with no possibility of action or existence beyond the brain. In conclusion, mediumistic experiences provide a large and diversified body of empirical evidence that strongly suggest a nonreductionist view of mind.

To improve the advancement of this exploration, it is necessary to have an investigative approach that in fact deserves being called *scientific*: a methodologically rigorous, epistemologically, and historically well informed, nondogmatic and open-minded approach. It is necessary for the academic environment, if it truly wants to understand human nature, does not exclude any sort of human experience, no matter how odd it may seem. In this phenomenological investigation, consistent empirical data need to have epistemological priority over established paradigms that are inadequate to explain many anomalous phenomena. To improve future research, it is necessary to have well-trained scientists, funding, and the scientific creativity to design new research protocols that are adequate to formulate and test theories that can explain the available data.

References

Aksakof, A. (1890/1994). Animismo e Espiritismo. Rio de Janeiro: FEB.

Almeder, R. (1992). *Death and personal survival: The evidence for life after death*. Lanham, MD: Rowman and Littlefield.

Almeida, A. A. S. (2007). *"Uma fábrica de loucos": Psiquiatria x espiritismo no Brasil (1900–1950). ["An insanity factory": Psychiatry and spiritism in Brazil]*. Doctoral dissertation, Department of History, Unicamp, Campinas. Retrieved September 16, 2011, from http://www.hoje.org.br/bves.

Almeida, A. M., & Lotufo Neto, F. (2004). A mediunidade vista por alguns pioneiros da área da saúde mental (Mediumship as seen by some pioneers in the area of mental health). *Revista de Psiquiatria Clínica, 31*, 132–141.

Alvarado, C. S. (2002). Dissociation in Britain during the late nineteenth century: The Society for Psychical Research, 1882–1900. *Journal of Trauma & Dissociation, 3*(2), 9–33.

Alvarado, C. S. (2003). The concept of survival of bodily death and the development of parapsychology. *Journal of the Society for Psychical Research, 67*, 65–95.

Alvarado, C. S. (2010). Investigating mental mediums: Research suggestions from the historical literature. *Journal of Scientific Exploration, 24*(2), 197–224.

Alvarado, C. S., Machado, F. R., Zangari, W., & Zingrone, N. L. (2007). Historical perspectives of the influence of mediuim ship on the construction of psychological and psychiatric ideas. *Revista de Psiquiatria Clínica, 34*(Supp. 1), 42–53.

Asprem, E. (2010). A nice arrangement of heterodoxies: William McDougall and the professionalization of psychical research. *Journal of the History of the Behavioral Sciences, 46*(2), 123–143.

Aubrée, M., & Laplantine, F. (1990). *La table, le livre et les esprits (The table, the book, and the spirits)*. Paris: Éditions Jean-Claude Làttes.

Beischel, J. (2007/2008). Contemporary methods used in laboratory-based mediumship research. *Journal of Parapsychology, 71*, 37–68.

Beischel, J., & Schwartz, G. E. (2007). Anomalous information reception by research mediums demonstrated using a novel triple-blind protocol. *Explore: The Journal of Science and Healing, 3*, 23–27.

Bem, D. J. (2005). Review of the afterlife experiments. *The Journal of Parapsychology, 69*, 173–183.

Berkeley, G. (1975). *Philosophical works*. In: MR Ayers (Ed.), London: Everyman.

Bourguignon, E. (1973). A framework for the comparative study of altered states of consciousness. In E. Bourguignon (Ed.), *Religion, altered states of consciousness, and social change* (pp. 3–38). Columbus: Ohio State University Press.

Bourguignon, E. (1976). *Possession*. San Francisco: Chandler & Sharp.

Bozzano, E. (1980). *Xenoglossia*. Rio de Janeiro: Federação Espírita Brasileira.

Braude, S. E. (2003). *Immortal remains: The evidence for life after death*. Lanham, MD: Rowman and Littlefield.

Brown, M. F. (1997). *The channeling zone: American spirituality in an anxious age*. Cambridge: Harvard University Press.

Cardeña, E., Lewis-Fernandez, R., Beahr, D., Pakianathan, I., & Spiegel, D. (1996). Dissociative disorders. In T. A. Widiger, A. J. Frances, H. J. Pincus, R. Ross, M. B. First, & W. W. Davis (Eds.), *Sourcebook for the DSM-IV (Vol II)* (pp. 973–1005). Washington, DC: American Psychiatric Press.

Carvalho, A. (1945). *Guerra Junqueiro e a sua obra poética*. Livraria Figueirinhas: Porto.

Chalmers, A. F. (1982). *What is this thing called Science?* (2nd ed.). Buckingham: Open University Press.

Chibeni, S. S., & Moreira-Almeida, A. (2007). Remarks on the scientific exploration of "anomalous" psychiatric phenomena. *Revista de Psiquiatria Clínica, 34*(Suppl. 1), 8–15. Retrieved September 16, 2011, from www.hoje.org.br/elsh.

Cook, E. F. W. (1992). *Frederic W. H. Myers: Parapsychology and its potential contribution to psychology*. PhD dissertation. University of Edinburgh, Edinburgh.

Crabtree, A. (2003). 'Automatism' and the emergence of dynamic psychiatry. *Journal of the History of the Behavioral Sciences, 39*(1), 51–70.

Crabtree, A. (1993). *From Mesmer to Freud: Magnetic sleep and the roots of psychological healing*. New Haven, CT: Yale University Press.

Crookes, W. (1874). *Researches in the phenomena of spiritualism*. London: J. Burns.

Delanne, G. (1898) *Recherches sur la Médiumnité*. J. Meyer (Ed.). Paris: BPS.

Ducasse, C. J. (1962). What would constitute conclusive evidence of survival after death? *Journal of the Society for Psychical Research, 41*(714), 401–406.

Ellenberger, H. F. (1970). *The discovery of the unconscious*. New York: Basic Books.

Eysenck, H. J., & Sargent, C. (1993). *Explaining the unexplained: Mysteries of the paranormal*. London: Prion.

Ferreira, J. M. H. (2004). *Estudando o invisível: William Crookes e a nova força*. São Paulo: EDUC/FAPESP.

Flammarion, C. (1900/1979). *O desconhecido e os problemas psíquicos (2 vol.)*. Rio de Janeiro: FEB.

Flournoy, T. (1900). *From India to the planet Mars: A study of a case of somnambulism*. New York: Harper & Brothers.

Gauld, A. (1982). *Mediumship and survival: A century of investigations*. London: Granada.

Gauld, A. (1968). *The founders of psychical research*. London: Routledge & Kegan Paul.

Hageman, J. H., Peres, J. F. P., Moreira-Almeida, A., Caixeta, L., Wickramasekera, I., II, & Krippner, S. (2010). In S. Krippner & H. Friedman (Eds.), *Mysterious minds: The neurobiology of psychics, mediums and other extraordinary people* (pp. 85–111). Santa Barbara, CA: Praeger.

Haraldsson, E., & Stevenson, I. (1975a). A communicator of the "drop in" type in Iceland: The case of Gudni Magnusson. *The Journal of the American Society for Psychical Research, 69*, 245–261.

Haraldsson, E., & Stevenson, I. (1975b). A communicator of the "drop in" type in Iceland: The case of Runolfur Runolfsson. *The Journal of the American Society for Psychical Research, 69*, 33–59.

Hastings, A. (1991). *With the tongues of men and angels: A study of channeling*. Fort Worth, TX: Holt, Rinehart and Winston.

Hodgson, R. (1892). A record of observations of certain phenomena of trance. *Proceedings of the Society for Psychical Research, 8*, 1–167.

Hodgson, R. (1898). A further record of observations of certain phenomena of trance. *Proceedings of the Society for Psychical Research, 13*, 284–582.

Hume, D. (2007). *A treatise of human nature (first published 1739–40)*. Oxford: Clarendon.

Hume, D. (2000). *An enquiry concerning human understanding (first published, 1748, as Philosophical essays concerning human understanding)*. Oxford: Clarendon.

Hyslop, J. H. (1905a). *Problem of philosophy or principles of epistemology and metaphysics*. New York: Macmillan.

Hyslop, J. H. (1905b). *Science and a future life*. Boston: Herbert B. Turner.

Inglis, B. (1982). Foreword. In A. Gauld (Ed.), *Mediumship and survival: A century of investigations* (pp. xi–xiv). London: Granada.

James, W. (1986). *Essays in psychical research*. Cambridge, MA: Harvard University Press.

James, W. (1886/1960). Report of the comittee on mediumistic phenomena. In: Murphy, G. & Ballou, R. O. (Eds.), *William James on psychical research* (pp. 95–100). New York: Viking.

James, W. (1890/1960). Certain phenomena of tranc. In: Murphy, G. & Ballou, R. O. (Eds.). *William James on psychical research* (pp. 102–111). New York: Viking.

James, W. (1896/1986). Address of the president before the society for psychical research. In: W. James (Ed.), *Essays in psychical research* (pp. 127–137). Cambridge, MA: Harvard University Press.

James, W. (1897/1960). *What psychical research has accomplished*. In: Murphy, G. & Ballou, R. O. (Ed.), William James on psychical research (pp. 25–47). New York: Viking.

James, W. (1898/1960). Human immortality: Two supposed objections to the doctrine. In: Murphy, G. & Ballou, R. O. (Ed.). *William James on psychical research* (pp. 279–308). New York: Viking

James, W. (1909/1960). Report on Mrs. Piper's Hodgson-control. In: Murphy, G. & Ballou, R. O. (Eds.), *William James on psychical research* (pp. 115–210). New York: Viking.

James, W. (1909a/1960). The final impressions of a psychical researcher. In: Murphy, G. & Ballou, R. O. (Eds.). *William James on psychical research* (pp. 309–325). New York: Viking.

Janet, P. (1889). *L'automatisme psychologique: Essai de psychologie expérimentale sur les forme inférieures de l'activité humaine*. Paris: Félix Alcan.

Jensen, C. G., & Cardeña, E. (2009). A controlled long-distance test of a professional medium. *European Journal of Parapsychology, 24*, 53–67.

Jung, C.G. (1983). On the psychology and pathology of so-called occult phenomena. In: *Psychiatric studies* (pp. 3–88). Princeton University Press, Princeton.

Kardec, A. (1986/1861) *The medium's book*. Rio de Janeiro: FEB.

Kardec, A. (1858a). Evocations particulières. Mère, je suis là! *Revue Spirite: Journal d'Études Psychologiques, 1*(1), 17–19.

Kardec, A. (1858b). Une leçon d'écriture par un Esprit. *Revue Spirite: Journal d'Études Psychologiques, 1*(7), 196–198.

Kardec, A. (1858c). Une nuit oubliée ou la sorcière Manouza. Dictée par l'Esprit de Frédéric Soulié. Préface De L'éditeur. *Revue Spirite: Journal d'Études Psychologiques, 1*(11), 315–317.

Kardec, A. (1859a). Dirkse Lammers. *Revue Spirite: Journal d'Études Psychologiques, 2*(12), 336–337.

Kardec, A. (1860). Etudes sur l'Esprit des personnes vivantes. Le Docteur Vignal. *Revue Spirite: Journal d'Études Psychologiques, 3*(3), 81–87.

Kardec, A. (1863). Examen des communications médianimiques qui nous sont adressées. *Revue Spirite: Journal d'Études Psychologiques, 6*(5), 156–159.

Kardec, A. (1859/1999). What is spiritism? Philadelphia: Allan Kardec Educational Society.

Kelly, E. W. (2010). Some directions for mediumship research. *Journal of Scientific Exploration, 24*, 247–282.

Kelly, E. W., & Arcangel, D. (2011). An investigation of mediums who claim to give information about deceased persons. *The Journal of Nervous and Mental Disease, 199*(1), 11–17.

Kelly, E. F., Kelly, E. W., Crabtree, A., Gauld, A., Grosso, M., & Greyson, B. (2007). *Irreducible mind: Toward a psychology for the 21st century*. Lanham: Rowman & Littlefield.

Klimo, J. (1998). *Channeling: Investigations on receiving information from paranormal sources*. Berkeley: North Atlantic Books.

Kottler, M. J. (1974). Alfred Russel Wallace, the origin of man, and spiritualism. *Isis, 65*(2), 144–192.

Krippner, S. (2007). Humanity's first healers: Psychological and psychiatric stances on shamans and shamanism. *Rev Psiquiatr Clín, 34*(Suppl 1), 17–24.

Lachapelle, S. (2002). *A world outside science. French attitudes toward mediumistic phenomena (1853–1931)*. Dissertation in history, University of Notre Dame, Notre Dame.

Lamont, P. (2004). Spiritualism and a mid-Victorian crisis of evidence. *The Historical Journal, 47*(4), 897–920.

Le Maléfan, P. (1999). *Folie et Spiritisme: Histoire du Discourse Psychopathologique sur la Pratique du Spiritisme, Ses Abords et Ses Avatars (1850–1950)*. Paris: L'Hartmattan.

Lester, D. (2005). *Is there life after death? An examination of the empirical evidence*. Jefferson, NC: McFarland.

Locke, J. (1975). *An essay concerning human understanding*. In: P. H. Nidditch (Ed.). Oxford: Clarendon.

Lodge, O. (1909). *The survival of man*. New York: Moffat, Yard.

Lombroso, C. (1909). *Richerche sui Fenomeni Ipinotici e Spiritici*. Torino: Fratelli Bocca.

Matos, M. V. (2011). Biografia de Chico Xavier. Federação Espírita Brasileira. Retrieved September 16, 2011, from http://www.100anoschicoxavier.com.br/paginas/biografia.html.

Monroe, J. W. (2008). *Laboratories of faith: Mesmerism, spiritism, and occultism in modern France*. Ithaca: Cornell University Press.

Moreira-Almeida, A. (2008). *Allan Kardec and the development of a research program in psychic experiences*. Parapsychological Association and The Society for Psychical Research Convention, Winchester, UK. Proceedings of Presented Papers 51, pp. 327–332.

Moreira-Almeida, A. & Cardeña, E. (2011). Differential diagnosis between non-pathological psychotic and spiritual experiences and mental disorders: A contribution from Latin American Studies to the ICD-11. *Revista Brasileira de Psiquiatria, 33*(suppl. 1), S29–S36.

Moreira-Almeida, A. (2006). Review of the book is there life after death? An examination of the empirical evidence, by David Lester. *Journal of Near-Death Studies, 24*, 245–254.

Moreira-Almeida, A., Almeida, A. A. S., & Lotufo Neto, F. (2005). History of spiritist madness in Brazil. *History of Psychiatry, 16*, 5–25.

Moreira-Almeida, A., Lotufo Neto, F., & Cardeña, E. (2008). Comparison of Brazilian spiritist mediumship and dissociative identity disorder. *The Journal of Nervous and Mental Disease, 196*, 420–424.

Moreira-Almeida, A., Lotufo Neto, F., & Greyson, B. (2007). Dissociative and psychotic experiences in Brazilian spiritist mediums. *Psychotherapy and Psychosomatics, 76*, 57–58. Erratum in: *Psychotherapy and Psychosomatics, 76*, 185.

Murphy, G., & Ballou, R. O. (1960). *William James on psychical research*. New York: Viking Press.

Myers, F. W. H. (2001/1903). *Human personality and its survival of bodily death*. Charlottesville: Hampton Roads.

Negro Jr, P. J. (1999). *A Natureza da Dissociação: Um Estudo sobre Experiências Dissociativas Associadas a Práticas Religiosas [tese]*. Faculdade de Medicina da Universidade de São Paulo, São Paulo.

Negro, Jr., P. J., Palladino-Negro, P., & Louzã, M. R. (2002). Do religious mediumship dissociative experiences conform to the sociocognitive theory of dissociation? *Journal of Trauma & Dissociation, 3*, 51–73.

O'Keeffe, C., & Wiseman, R. (2005). Testing alleged mediumship: Methods and results. *British Journal of Psychology, 96*, 165–179.

Owen, A. (2004). *The place of enchantment: British occultism and the culture of the modern*. Chicago: Chicago University Press.

Perandréa, C. A. (1991). *A Psicografia à luz da grafoscopia*. São Paulo: Editora Jornalística Fé.

Perry, J. (2008). *Personal identity*. Berkeley: University of California Press.

Popper, K. (1963). *Conjectures and refutations*. London: Routledge.

Quinton, A. (1962/2008). The soul. In: Perry, J. (Ed.), *Personal identity* (pp. 53–72). Berkeley: University of California Press.

Raia, C. G. (2007). From ether theory to ether theology? Oliver Lodge and the physics of immortality. *Journal of the History of the Behavioral Sciences, 43*(1), 19–43.

Ravaldini, S., Biondi, M., & Stevenson, I. (1990). The case of Giuseppe Riccardi: An unusual drop-in communicator in Italy. *Journal of the Society for Psychical Research, 56*(821), 257–265.

Rhine, J. B. (1937). *New frontiers of the mind*. New York: Farrar & Rinehart.

Rhine, J. B. (1956). Research on spirit survival re-examined. *The Journal of Parapsychology, 20*(2), 121–131.

Robertson, T. J., & Roy, A. E. (2001). A preliminary study of the acceptance by non-recipients of medium's statement to recipients. *Journal of the Society for Psychical Research, 65*, 91–106.

Rocha, A. C. (2001a). A poesia transcendente de Parnaso de além-túmulo. Tese (mestrado). Universidade Estadual de Campinas, Instituto de Estudos da Linguagem, Campinas. Retrieved September 16, 2011, from www.hoje.org.br/bves.

Rocha, A. C. (2001b). Guerra Junqueiro de Chico Xavier. EPA – Estudos portugueses e Africanos 38 (Jul/Dez), pp. 119–141.

Rocha, A. C. (2008). O Caso Humberto De Campos: Autoria Literária e Mediunidade. Tese de Doutorado em Teoria e História Literária, Unicamp. Retrieved September 16, 2011, from www.hoje.org.br/bves.

Roy, A. E., & Robertson, T. J. (2001). A double-blind procedure for assessing the relevance of a medium's statements to a recipient. *Journal of the Society for Psychical Research, 65*, 161–174.

Roy, A. E., & Robertson, T. J. (2004). Results of the application of the Robertson-Roy protocol to a series of experiments with mediums and participants. *Journal of the Society for Psychical Research, 68*, 18–34.

Schwartz, G. E., Geoffrion, S., Jain, S., Lewis, S., & Russek, L. G. (2003). Evidence of anomalous information retrieval between two mediums: Replication in a double-blind design. *J Soc Psych Res., 67*, 115–130.

Schwartz, G. E. R., Russek, L. G. S., & Barentsen, C. (2002). Accuracy and replicability of anomalous information retrieval: Replication and extension. *Journal of the Society for Psychical Research., 66*(3), 144–156.

SDRJ. (2011). Ranking 2010 Brasil. Retrieved September 16, 2011, from www.filmeb.com.br.

Severino, P. R. (1994). *Life's triumph*: research on messages received by Chico Xavier. São Paulo: Jornalística Fé.

Sharp, L. L. (2006). *Secular spirituality: Reincarnation and spiritism in nineteenth-century France*. Lanham: Lexington Books.

Sidgwick, E. M. (1915). *A contribution to the study of the psychology of Mrs. Piper's trance phenomena*. Proceedings of the Society for Psychical Research, *28*, pp. 1–657.

Souto Maior, M. (2003). *As vidas de Chico Xavier* (2nd ed.). São Paulo: Editora Planeta.

Stevenson, I. (1970). A communicator unknown to medium and sitters. *The Journal of the American Society for Psychical Research, 64*, 53–65.

Stevenson, I. (1973). A communicator of the "drop in" type in France: The case of Robert Marie. *The Journal of the American Society for Psychical Research, 67*(1), 47–76.

Stevenson, I. (1974). *Xenoglossy – A review and report of a case*. Bristol: John Wright.

Stevenson, I. (1976). A preliminary report of a new case of responsive xenoglossy: The case of Gretchen. *The Journal of the American Society for Psychical Research, 70*, 65–77.

Stevenson, I. (1977). Research into the evidence of man's survival after death: A historical and critical survey with a summary of recent developments. *The Journal of Nervous and Mental Disease, 165*(3), 152–170.

Stevenson, I. (1983). Do we need a new word to supplement "hallucination"? *The American Journal of Psychiatry, 140*(12), 1609–1611.

Stevenson, I. (2007). Half a career with the paranormal. *Revista de Psiquiatria Clínica., 34*(Suppl. 1), 150–155.

Stevenson, I., Pasricha, S., & McClean-Rice, N. (1989). A case of the possession type in India with evidence of paranormal knowledge. *Journal of Scientific Exploration, 3*(1), 81–101.

Stromberg, W. H. (1989). Helmholtz and Zoellner: Nineteenth-century empiricism, spiritism, and the theory of space perception. *Journal of the History of the Behavioral Sciences, 23*, 371–383.

Thuillier, P. (1977). Evolutionnisme et spiritisme: Le cas Wallace. *La Recherche, 80*(8), 691–696.

Treitel, C. (2004). *A science for the soul: Occultism and the genesis of the German Modern*. Baltimore: Johns Hopkins University Press.

Webster, M. (1996). *Webster's comprehensive dictionary of the English language*. Naples: Trident.

Chapter 11
Cases of the Reincarnation Type and the Mind–Brain Relationship

Erlendur Haraldsson

Abstract Children who claim to have memories of a previous life are sometimes found, particularly in countries with widespread belief in reincarnation. Over 2500 such cases have been investigated in many countries, including some in the USA and Europe. Analysis of the contents of these alleged memories reveals interesting features, such as frequent images/memories of the mode of death which in most cases is due to accidents or other violent means. Phobias related to the mode of death are common and birthmarks and deformities are sometimes found that the child relates to the way he or she died. Children start to speak about these images/ memories almost as soon they can speak, usually around two and half to three years. Psychological studies reveal interesting differences between them and their peers. Attempts to verify these memories have had some degree of success as a deceased person has in some instances been found whose life events correspond to the child´s statements. In other instances verification has failed. Four cases are presented from Lebanon and Sri Lanka. If these alleged memories are genuine they have great implications for the mind-brain relationship.

11.1 Introduction

Children who claim to remember episodes from a past life have been found in many countries. The genuineness of their alleged memories has been a matter of considerable discussion as they are potentially highly relevant for the question of the mind–brain relationship. If these are real memories from a past life they indicate that memory is not only stored in the brain but also that mind can exist without a brain and still retain some of its memories. These possibilities are radically contrary to what is presently known about memory and its dependence on brain functioning,

E. Haraldsson (✉)
Department of Psychology, University of Iceland, 101, Reykjavik, Iceland
e-mail: Erlendur@hi.is

A. Moreira-Almeida and F.S. Santos (eds.), *Exploring Frontiers of the Mind-Brain Relationship*, Mindfulness in Behavioral Health, DOI 10.1007/978-1-4614-0647-1_11,
© Springer Science+Business Media, LLC 2012

and hence meet with extreme skepticism by the scientific community. This chapter will describe the main characteristics of these cases, how they are investigated, and four cases will be presented to reveal to the reader their strengths and weaknesses. For a general introduction to past-life memories see Haraldsson (2001), Stevenson (2001), and Tucker (2005).

Cases of children claiming past-life memories are sometimes discovered by the media and lead to headlines, particularly in countries where belief in reincarnation is widespread, but most of them remain unknown to the public. Often the parents of the child make an effort to keep the case within the family and only some relative or researcher is allowed to examine the case. Half a century ago, Professor Ian Stevenson of the University of Virginia, United States, conducted the first systematic study of *cases of the reincarnation type* as termed them (Stevenson 1960, 1997a, b, 2001). Since then over 2,500 cases have been recorded and investigated worldwide, most of them by Stevenson. Files on the cases are kept at the Division of Perceptual Studies at the University of Virginia.

At the end of the 1980s, Stevenson asked the author if he would be willing to make an independent study of such cases. Would a study of new cases by an independent researcher reveal findings comparable to those of Stevenson? Since that time the author has investigated over 60 cases in Sri Lanka (Haraldsson 1991, 2000a, b; Haraldsson and Samararatne 1999), 30 in Lebanon (Haraldsson and Abu-Izzedin 2002, 2004) and a handful in other countries, among them one in Iceland. He has conducted three psychological studies comparing these children with peers with no claims of past-life memories (Haraldsson 1997, 2003; Haraldsson et al. 2000).

Cases of the reincarnation type are rare. In Sri Lanka, the author was able to find about five per year in a population of 18 millions. Most of them were among Buddhist families and a few in Christian (6) and Muslim families (3). In Lebanon, the cases are somewhat easier to find but practically only within the Druze community where belief in reincarnation is widespread. However, even there the author learned of two cases in the Christian community, one of them investigated by Stevenson in the 1970s.

Regarding the content of these alleged memories (for brevity sake from now on referred to as memories only), most children speak of how they died, which is in most cases violently through accidents, murder, or acts of war. They speak of events that lead to their death, of people they knew, and where they had lived. They speak of recent events occurring in or near the area where they lived. There is also a behavioral aspect as many suffer from phobias, which they associate with the mode of death they described.

In most cases, children start to speak of a past life around the age of two and half to 3, or almost as soon as they can speak. The numbers of statements they make varies considerably and are on the average around 20. These children commonly request that their parents find their previous home, and say that their present mother is not their real mother. In most instances, the children stop to talk about their past life around the time they go to school. Occasionally, they reveal knowledge or skills that they were not known to have learnt, and some have birthmarks or deformities that they relate to the mode of death in the previous life. Psychological studies show characteristics that reveal phobias related to their memories, and symptoms of a posttraumatic stress disorder (PTSD) in these children as a group, but not necessarily in each one of them (Haraldsson 2003). The most likely explanation for the PTSD symptoms is the memories of a violent death that up to 75% of these children report in early childhood. Any explanatory theory about the nature of these cases has to accommodate not only the memory aspect but also the birthmarks, the psychological characteristics, and the PTSD.

11.2 Methods of Investigation

The first step is to interview the child, his/her parents, siblings, and other persons who may have observed the child talking about the previous life. This is to ascertain what statements the child has made and if s/he has done that consistently on several occasions. The second step is to rule out that the child is talking about events that s/he has learned about from his/her environment. Only if that is not the case and there is reasonable consensus about the statements made by the child, will the case be considered worthy of further investigation. It is important to interview the principal witnesses again after some time to test the reliability of their testimony.

The next question is if a deceased person can be traced whose life events correspond to the statements made by the child. Often a person has been found who the child is believed to be referring to. Such a correspondence has to be checked to see if it does in fact exist. The family of that person has to be interviewed and relevant documents obtained, such as birth and death certificates, postmortem reports, etc. as the case may be. There follows a detailed review of four cases that were investigated by the author.

11.2.1 The Case of Purnima Ekanayaka

Later visit to the temple, she claimed that she had lived on the other side of the Kelaniya Purnima Ekanayaka was 9 years old when we met her in 1996. She was still speaking of her previous life and made a fairly large number of statements about it. She had been talking about her previous life since she was 3 years old. Her case had been *solved*, that is, a deceased person had been identified and accepted as her previous personality. Her case had the weakness that it had been published in the newspapers, so it was difficult to know for sure what Purnima´s original statements were. In Table 11.1 are listed statements which, according her and her parents, she made before she met the family that came to be accepted as her previous family.

One of Purnima's first statements was that she had died in a traffic accident and "came here," namely was reborn in her present family. "My family was making incense," and she stated. The brands were "Ambika" and "Gita Pichcha." Their incense factory was near a brick factory and near a pond. Furthermore, she stated that first only the family worked in the incense business, and then some people were employed. She had two vans, and a car, was the best manufacturer of the incense sticks, was married and had a sister-in-law. She spoke of herself as a man in her previous life and he/she had two wives. She said that the previous father had not

Table 11.1 Statements made by Purnima (according to her parents) before first contact with her alleged previous family

I died in a traffic accident and came here	+
My family was making incense and had no other job	+
We were making Ambiga incense	+
We were making Geta Pichcha incense	+
The incense factory is near a brick factory and near a pond	+
First only our family worked and then two people were employed	?
We had two vans	
We had a car	
I was the best manufacturer of incense sticks	?
In earlier birth I was married to a sister-in-law, Kusumi	+
The owner of the incense factory (I) had two wives	+
My previous father was bad (present father is good)	?
Previous father was not a teacher as present father	+
I had two younger brothers (who were better than present brothers)	+
My mother's name was Simona	+
Simona was very fair	−
I attended Rahula School	+
Rahula School had a two storied building (not like in Bakamuna)	−
My father said, you need not go to school, you can make money making incense	
I studied only up to fifth grade	+

+, Verified (14); −, incorrect (3); ?, indeterminate (3)

been as good as the present one, and was not a teacher like he is (he is a principal of a secondary school). One evening when the family saw a documentary on the famous Kelaniya temple on television, she said that she recognized the temple. During a River which flows alongside the temple.

In January 1993, a graduate of Kelaniya University, W.G. Sumanasiri, was appointed a teacher in Bakamuna, but spent the weekends with his wife in Kelaniya which is 145 miles away. Sumanasiri and Purnima´s father agreed that he would make inquiries across the Kelaniya River. The principal gave him the following items to check:

- She had lived on the other side of the river from Kelaniya temple.
- She had been making Ambika and Gita Pichcha incense sticks.
- She was selling incense sticks on a bicycle.
- She had a fatal accident with a big vehicle.

Sumanasiri and his brother-in-law inquired about incense makers in the area, and were told of L.A. Wijisiri who named his brands Ambika and Gita Pichcha. His brother-in-law and associate, Jinadasa Perera, had died in an accident with a bus as he was bringing incense to the market on a bicycle in September 1985. This was about 2 years prior to Purnima's birth. Wijesiri's and Jinadasa's home and factory had been 5–10 min walking distance from the Kelaniya River, and 2.4 miles away from the temple. Wijesiri's wife is the sister of Jinadasa who died in the accident. Earlier Wijesiri and Jinadasa had run this business together.

Of the 20 statements listed in Table 11.1, there are 14 that correctly fit the life of Jinadasa. Among them the circumstances and mode of death, occupation, and names of the brands of incense that they and only they produced, and description of the vicinity of their factory. Three statements were incorrect. The school that Jinadasa attended did not have two stories until after Jinadasa had left school. Also, it cannot be correct that Jinadasa's father had said that Jinadasa should not continue to go to school in order to make incense because Jinadasa did not start to make incense until his sister married Wijesiri and he got involved with the family around the age of 20. Three statements are indeterminate, and of little relevance. Witnesses claim that when Purnima visited the family of Wijesiri for the first time she recognized two old friends.

Purnima had from her birth a prominent birthmark, a cluster of hypopigmented spots over the ribs on the left side of her chest. Jinadasa had died in a traffic accident and the wheel of the bus had run over him and broke many of his ribs. The postmortem report for Jinadasa revealed that many of his ribs had broken on the left side of his chest. They penetrated the lungs and in this is the area he is likely to have suffered most. The spleen and the liver were ruptured. Purnima's birthmarks are on the left side of her chest and fit the location of the principal injuries on the body of Jinadasa. The birthmarks are an additional characteristic that falls in place with the life of Jinadasa. Cases of birthmarks that fit the wounds of the alleged previous person are particularly interesting as they are already formed when the child is born. The author has investigated another birthmark case in Sri Lanka, that of Chatura Karunaratne (Haraldsson 2000b). That case has the additional feature, like the case of Thusita Silva to be described below, that the statements by the boy were recorded and printed before a fitting previous person was found.

There is one rare characteristic of this case. Purnima spoke of her life between death in the accident and being born in Bakamuna. She claims to have observed what took place after she died, at the place of the accident and who was present at the funeral. Then she saw a light, went there, and was born in Bakamuna. Jinadasa Perera was in born in 1949 and he died in April 1985. Purnima was born in August 1987 and the case is solved in 1993.

Purnima spoke quite vividly and in detail about her memories. Like many children who speak about a previous life, psychological tests administered to her, revealed that she had excellent memory, and scored high on a brief test of intelligence. She is on the top of a class of 23 pupils, a bright and gifted girl. For further details see the full report on the case (Haraldsson 2000a, b).

This case has weaknesses that are common among cases of this kind. We did not interview Purnima about her memories until after she had met the Wijisiri family. The question remains if some of her memories did not fit the life of Jinadasa and were consequently suppressed, forgotten, or distorted to better fit the life of Jinadasa. However, the fact remains that she spoke of being an incense maker, having died in an accident with a "big" vehicle and having lived across the Kelaniya River near the temple. There is also a remarkable fit between the mortal wounds of Jinadasa and the location of Purnima's extensive birthmarks. The two families were complete strangers and lived far apart. Of some interest fact that both families became convinced that Purnima was Jinadasa reborn, because they witnessed the verification of her memories and some recognition of persons Jinadasa had known.

The following case is free of the principal weaknesses of these cases that were described above.

11.2.2   The Case of Thusita Silva

This case concerns a girl we shall call Thusitha Silva (pseudomyn). It is particularly interesting because the girl´s statements were recorded by us before any person was found whose life events corresponded to the events that the girl had been talking about. When we met her in 1991 she was 8 years old and lived in poor conditions near the town of Panadura, which is south of Colombo in Sri Lanka. When Thusita had been talking about her previous life for a while her much older only brother went to Akuressa (population about 20.000) to check on her story, failed to verify it, and scolded the girl. No further attempts were made to solve the case. A few years had passed when we learned about the case and interviewed the girl, her mother, and grandmother. By this time she had forgotten some of her earlier memories.

According to her mother and grandmother, she had at the age of two and half claimed; "I am from Akuressa, my father's name is Jeedin Nanayakkara," furthermore that she had lived near a river, and when crossing a narrow footbridge she fell into the river and drowned, pregnant at the time. She had a husband and the house she had lived in was larger than the present house. A list of her statements is in Table 11.2.

During a visit to Akuressa we found a Nanayakkara family that lived near a hanging bridge for pedestrians. We met the family who told us that their daughter-in-law had been crossing the bridge in 1973 when she fell off the bridge and drowned. Her husband was with her and jumped into the river to save her but almost drowned himself. There was a file on the case in the coroner´s office in Akuressa. Chandra Nanayakkara (born Abeygunasekera) had died in December 1973, by choking after swallowing water when the deceased fell into the River Nilwala from the suspension bridge. "She was at the age of 27, and 7 months pregnant. She had been with her husband when the accident occurred."

Seventeen of Thusitha's 28 statements fitted the life of Chandra Nanayakkara; seven were incorrect and four indeterminate. All of Thusita´s statements relating to the mode of death fitted, namely, the hanging bridge, falling into the river, that she was pregnant, and that her husband was with her. The name of her father was not Jeedin Nanayakkara but her father-in-law was Edwin Nanayakkara. Women in Sri Lanka often refer to their father-in-law as father. The incorrect statements were about the color of her bicycle, that she had worked in a hospital (her best friend did) and that her husband was a postman (his brother was) but he was a bus driver. Other statements were too general to be of much value (Mills et al. 1994).

Thusita´s family claimed to have no connection of any kind with Akuressa and none of them had been there when Thusita spoke most about her previous life. Akuressa is about 50 km away from Thusita's birthplace in Elpitiya from where she moved to Panadura where we met her.

Cases in which the child´s statements were recorded before any previous person was identified are of great importance since we can then be sure that the child´s statements are uncontaminated. Further recent cases of this kind are the cases of Chatura Karunaratne (Haraldsson 2000b), Dilukshi Nissanka (Haraldsson 1991),

Table 11.2 Thusita Silva´s statements about her previous life

I lived in Akuressa	+
My father´s name was Jeedin	−
(my father´s name was) Nanayakkara	+
River or stream a little distance away	+
The hanging bridge (wel palama) broke down	+
I fell into the river	+
I drowned	+
I was pregnant when drowned	+
I had a husband	+
Our house was larger than present house	+
Walls were colored	+
I had a sister´s daughter	−
My former father was called appa (present father dada)	?
I had a bicycle	+
Bicycle was yellow	−
I went to work by bicycle	−
I rode the bicycle alone	+
I worked in a hospital	−
I wore a white uniform in hospital with cap and shoes	
Hospital was some distance from home	+
Mother wore frocks	?
Mother had a sewing machine	+
I had two striped frocks	?
Items reported to T.J. but not E.H.	
Big gate at former house	+
My husband jumped into the river to save me	+
My husband was a postman	−
We had a car	+
I had a brassiere	?

17 correct statements (+); 7 incorrect statements (−); 4 indeterminate (?)
The statements that fit the life of Chandra Nanayakkara are marked with
"+" those that do not fit with "−" and indeterminate statements with "?"

and cases investigated by Keil and Tucker (2005), Stevenson and Schouten (1998), and Stevenson and Samararatne (1988).

11.2.3   The Case of Prethiba Gunawardana

Prethiba Gunawardana was born in 1985 and was 4 years old when we met him and his mother in 1989 at their home in one of the suburbs of Colombo. My interpreter had learned about this case from a friend. It was only after we convinced Prethiba's mother that we would not publicize the case in Sri Lanka that she was willing to talk to us.

Prethiba had made his first statements about a previous life after he suffered high fever for a week when he was a little over 2 years old. After that he frequently spoke about his memories. He spoke to us without shyness about his memories that he insists are from a previous life. Pretiba is strongly built and healthy looking.

Pretiba stated that he had lived in Kandy (using the Sinhalese name, Maha Nuwara), the main city of central Sri Lanka. He gave his former name as Santha Megahathenne, and said that he had lived at number 28 Pilagoda Road. His car had caught fire, he had been burnt on his right leg, hand, and mouth, had been taken to a hospital and then he "came here" (died). His mother told us that he mentioned especially often two names: an older brother Samantha and an older sister Seetha. His father later told us that Pretiba often said he wanted to see them. According to his mother, he talked more often about names than events. His 42 statements are listed in Table 11.3. Pretiba appears to have no unusual behavioral traits that seem related to his statements. When we asked the boy if he would like to go to Kandy, he was quick to say yes. He said he could find his house, but when we asked him if he knew its

Whereabouts he replied with no. His father had not been willing to search for the previous personality, and the mother shared the common fear of mothers of such children that she might lose her child to the previous family if they were found. The boy had also told his parents that he wanted to go to Kandy to collect his things.

In Kandy, we made inquiries about Pilagoda Road and names resembling it. Post office authorities told us that there was no such road in Kandy city nor any village or area by that name in the Kandy district. We also made inquiries about the name Megahathenne, which Pretiba gave as his former family name. Some Sri Lankans use the name of the village they come from as a family name. A village by the name of Megahathenne exists near Galagedara some 15 miles away from Kandy. Inquiries there yielded no information about any person having the characteristics described by Pretiba, and no Pilagoda Road was found in that village, nor was the name Megahathenne found in the 1975 telephone directory for Kandy. The parents accepted our proposal to take the boy and them to Kandy.

The 3-h drive up the scenic road leads through many villages and towns. As we were approaching the bridge over the Mahaveli river at the other side of which is Kandy city, and were driving through a busy street, the boy became quite animated. He spontaneously said, "There is Maha Nuwara," (Sinhalese name for Kandy) and as we crossed the bridge (one of a few on the way) over the Mahaveli river, he correctly remarked, "This is Mahaveli Ganga" (Ganga = river). Neither we nor his parents had mentioned this name nor given any indication that we were about to enter Kandy city. Apart from these two statements, there was no response or comment from Pretiba to indicate any recognition or knowledge of the area. We drove for a while around Kandy but Pretiba could not tell us how to find his old home and expressed no wish to see a particular spot, although he definitely enjoyed the journey.

Enquiries in Kandy and Megahathenne failed to trace a person that fitted Pretiba's statements. It was impossible for us to go through the thousands of admissions every year to the Kandy hospital in the hope of finding the name of Santha Megahathenne. Without revealing the boy's name or address the main features of the case were publicized with the parent's permission in an interview with the author on December 11, 1990, in the widely circulated Singalese Dinamina and its English edition Daily News.

Table 11.3   Statements made by Pretiba Gunawardana about his previous life

Often mentions Samantha aya (elder brother)
Often mentions Seetha akka (elder sister)
Elder sister was married
Mentions Loku aya and Loku akka (big/elder)
Mentioned Dhamman Sadhu, a relative of father's brother
They had a car and a bus
His car had been burned (with much smoke) with him in it
Right hand, leg, and mouth had been burned
Admitted to Nuwara hospital, plaster placed on his body
After that he came to this place (died and was born here)
He had been to India and to a Hindu temple (kovil)
He had a passport
Mentioned name of Natapati (Nathapathi), visited Natapati Devalaya (kovil) while in India
Brought from India some items for his mother (saris and buttons)
He lived at number 28 Pilagoda Road in Nuwara (Kandy)
He lived upstairs in a house
His father was old
His father had a car
His father wore eyeglasses
Father had gone abroad and returned
Mentions a fight between snake and katussa
He had a girlfriend but did not like to marry that girl
They had a house with land around it
He had an uncle
They had paddyfields
Balansena worked in the paddyfields
There was a temple near the house
Artworks of elephants at the temple
He went to Sunday temple school
They had a refrigerator
They had a pettagama (large wooden box)
He had a good wristwatch
Mentions punchi amma (mother's younger sister)
Punchi amma's husband had a lorry and was a businessman
His name was Santha Megahathenne
He had a friend called Asanga
Bandara also lived there
He wore trousers
He was attending school
They had a bank account
His (former) brother looked like the brother of his (present) mother
Attanayake lived close to our house and had a lorry

No response came from readers. Despite these efforts to solve the case no person was found corresponding to Pretiba's statements. This case remains unsolved.

The last case to be described here is from Lebanon (Haraldsson and Abu-Izzedin 2004). It has some of the common weaknesses of these case but also contains some

Table 11.4 Statements made by Wael Kiwan according to his parents

My name was Rabih	+
I was big (not small)	+
I have parents. They are not here. They are in Beirut	+
My house is in Beirut near the sea	+
My house is near the house of Allah Wa Akbar (mosque)	+
There is a house with a red roof	+
It was sunset and I saw people coming and they shot me	−
A group of people hit me and kicked me until I did not feel anything	+
I was often on a boat out at sea	+
I used to stand and steer the boat with a wheel	?
I would walk from my house to the sea	+
My house is in Jal al Bahr	+
I had two homes, one in Beirut and one to which I go with an airplane	+
We had a balcony	+
I used to jump from the balcony to the street	+
I used to throw an "iron" to stop the boat (only reported by his aunt)	?
My (previous) mother is prettier than you	?
Twelve verified statements, two incorrect and three indeterminate	
The statements that fit the life of Rabih are marked with "+" those that do not with "−" and indeterminate statements with "?"	

highly specific statements made by the subject that made it possible to identify a deceased person whose life events closely resembled those that the child had described.

11.2.4 The Case of Wael Kiwan

Wael Kiwan lived in a village 70 km east of Beirut. He looked a healthy and mature boy when we met. His parents reported that at a young age he started to say that his name was Rabih, that he had been a grown-up person, had other parents in Beirut and wanted to find them. There were further statements: "There was a house with a red brick roof," and he lived in the Jal al Bahr section of Beirut (that is by the sea) and near a house of Allah Wa Akbar [i. e., a mosque], spoke much about the sea and a boat. He would draw a wheel of a boat on paper and say: "I used to stand," and he did a circular movement of his hands to show how he moved the steering wheel. He said that they had a balcony, from which he used to jump to the street. He made the highly specific statement that he had two homes, one in Beirut, and another one to which he had to travel by airplane.

He did what many children who speak of a previous life do; he said to his mother: "My [previous] mother is prettier than you." He often repeated the story about his death; it was sunset; he saw people coming towards him, and they shot him. When his father went on business to Beirut Wael would ask him to find his home. Wael would also tell him: "If you find it, don't tell them that Rabih has died, because they will cry." His statements are listed in Table 11.4. Listed in Table 11.4 are the statements that he, according to his parents, made before they started to seriously look for a deceased person that fitted his account.

Once his father, wife, and children sat down with Wael and listed many family names in the hope that he might recognize one of them. When they mentioned the name Assaf he said that was his previous family name. Assaf is carried by Druze, Christians, and Muslims alike and is a rather common name.

In vain Wael often asked his father to find his previous home when he went to Beirut on business. The boy would get upset when he returned and had not tried. Finally, he told a Druze friend in Beirut, Sami Zhairi, what Wael was saying and he promised to ask around. He found out that there had been a Rabih Assaf living close to the sea in the Jal-al-Bahr district whose life seemed to fit Weal's statements.

Wael was taken by his father to Beirut. Accompanied by Sami Zhairi they went to the house in the Jal al Bahr section. Wael ran into the house ahead of the group, into the apartment on the ground floor, where he saw a picture on the wall and said, "This is my picture." It was a picture of Rabih Assaf. In the apartment was Raja Assaf, the brother of the deceased Rabih (their mother, Munira, was not at home). Raja brought out a photo album and asked Wael to identify people. According to Wael's father, he recognized Rabin's father, sister, and a paternal aunt. When they left the house and were driving back home Wael told his father that now he was relaxed that he had found his previous home. Occasional visits followed and the families kept in contact over the years. After the meeting with the Assaf family, Wael spoke less about his former life.

Munira Assaf was at home when Wael came the second time. He seemed happy to be there but did not recognize Rabin's mother, his twin sister, or brother. Munira did not recall that he recognized any photographs, only that he said: "Yes, yes." She took this to mean that he might know the people in the photographs.

When Wael first visited the home of the Assaf family, he went to the backdoor and asked about the house with a red roof. This more than anything else made the family believe Wael was Rabih reborn; Rabih had grown up seeing this house from the kitchen and backyard of the apartment but it had been torn down when Wael visited them.

Munira Assaf's apartment is on the ground floor and has a balcony from which it is easy for a boy to jump to the street. She verified that Rabih had often done that. Wael had repeatedly mentioned a boat. Some 30 m down and near the end of the short street a small harbor is now crammed between houses and a huge high-rise apartment building. Munira's husband and sons had no boat but friends, relatives and neighbors did, and Rabih used to go to sea with them. Most of the boats were rowing boats, and a few had motors and had a tiller. These small boats did not have steering wheels. However, Rabih might have gotten a ride on a boat with a steering wheel. An old mosque is approximately 100 m away from Rabin's house on the same street as the harbor, and is the only mosque in the Jal al Bahr area.

The statement that Rabih had two homes, one of which you had to go to with an airplane, fits the fact that Rabih also lived in the United States. Item 16, "I used to throw an iron to stop the boat" was only reported to us by Weal's aunt Fadia. By iron ("hadideh" in Arabic) Wael probably meant an anchor, a word unlikely to be known by someone living far from the sea.

Rabin's mother expected to learn more from him about his life as Rabih but she did not. However, she was still convinced and accepted him, but also said: "Nothing will bring my son back."

Weal's mother told us that Wael had given two accounts of his death. The first was that "they" shot him in his head. The second version was that a group of people kicked him and hit him until he did not feel anything (hence his parents assumed that he had been killed that way).

Rabih died in South Pasadena, California in 1988. He had moved to the USA when he was 21 years old, and studied electrical engineering for 2 years. During the 3rd year, he wanted to return to Beirut, but was unable to do so because of the civil war in Lebanon, nor did he have enough money to either stay in California, or to return to Beirut. He was depressed, attempted suicide by swallowing pills, and was brought to a hospital and survived. Rabih moved to distant relatives. On January 9, 1988 his paternal cousin, Abboud Assaf, found him dead in his garage. He had hanged himself. This was verified in a telephone interview with him, nor was he aware of that a group of people kicked him and hit him until he did not feel anything. Also Rabin's mother did not know of any such incident.

One of the merits of the case is that the two families lived far apart and were complete strangers. The principal weakness is that the subject's statements were not recorded until after the two families had met. Twelve of the 17 statements Wael made correspond to events in the life of Rabih Assaf. Four items could neither be confirmed nor refuted. The statement that Rabih would stand in the boat holding a steering wheel, may have happened, but was certainly not the rule because the small boats on which he often went to sea would most likely not have had steering wheels.

One crucial item does not fit, namely the mode of death. This is the only major discrepancy between Weal's statements and the life of Rabih. Rabih committed suicide, whereas Wael speaks of being shot. If we allow some speculation and fantasy, then the following statement by Wael may carry some significance, "If you find my family, do not tell them that Rabih has died. They will cry." Could it possibly mean that he had some guilt about his mode of death or that he did not want it to be known? In view of the hanging the "announcing dream" of Wael's mother is also interesting, as she dreamt before Wael's birth a grown boy who was sweating, and breathing rapidly with difficulty. Could this be related to Rabih's hanging?

What corresponds well is the name Rabih, the family name Assaf, that he lived close to the sea and went to sea on boat(s), that there was a mosque close to his home, that he would jump from his balcony to the street, had lived in two places, to one of which he had to travel by airplane, and the statement that he made when he first visited the Assaf family, namely that behind his home there had been a house with a red roof. Many of these statements are highly specific and unlikely to be due to chance.

11.3   Psychological Characteristics of Children with Past-Life Memories

Two studies in Sr Lanka (Haraldsson 1997; Haraldsson et al. 2000) and one in Lebanon (Haraldsson 2003) show that children claiming past-life memories differ psychologically from other children. Stevenson noticed early on that they frequently suffered from phobias that seemed related to their past-life memories. Psychological studies comparing them with peers of the same age and from the same social background verified his clinical observation in Sri Lanka as well as in Lebanon. Results from two items on the Child Behaviour Checklist can be seen in Table 11.5.

Findings from the Childbehavior Checklist (CBCL) and the Child Dissociation Checklist (CDC) reveal symptoms that characterize PTSD patients with an identifiable trauma, such as phobias, fears, outbursts of anger, and nightmares. In both Sri Lanka and Lebanon, significant differences in the Problem score on the CBCL and an elevated CDC score indicate that children with past-life memories are often traumatized (see Table 11.6). Child abuse is one potential cause but there are no signs of abuse in their short life. Can the cause of the trauma be found in the images/

Table 11.5 Phobias and fears in subjects and controls in two combined studies in Sri Lanka. Results of two items in the child behavior checklist

	Subjects ($n = 57$)	Controls ($n = 57$)
Fear situations, places		
Often	16	4
Sometimes	16	18
Never	21	33
Too fearful or anxious		
Often	12	2
Sometimes	14	8
Never	28	45

Table 11.6 Psychological differences between subjects and controls in three studies in Sri Lanka and Lebanon

	Subjects	Controls	Z value
Child Behaviour Checklist, Problem score			
First Sri Lanka Study	41.33	26.77	3.73**
Second Sri Lanka Study	34.54	17.54	3.80**
Lebanon Study	45.10	27.70	3.73**
Child Dissociation Checklist			
Second Sri Lanka Study	6.59	1.69	3.80**
Lebanon Study	1.47	0.23	2.61*
Correlation between CBCL and CDC is 0.57 ($p < 0.001$)			
First Sri Lanka Study	Mean age, 9.39 years	12 boys	18 girls
Second Sri Lanka Study	Mean age, 7.83	14 boys	13 girls
Lebanon Study	Mean age, 10.62	19 boys	11 girls

memories of a past life that they report? That cause becomes more understandable when we consider that 77% of the Lebanon sample and 76% of the combined Sri Lanka samples speak of experiences of a sudden violent death. Further strengthening the PTSD argument is the finding that the Problem score for the CBCL is higher for children speaking of violent death (47.38, $N=24$) than for those who do not (36.00, $N=6$). The difference is significant ($t=2.43$, $p=0.03$, two-tailed).

To recapitulate, a high percentage of children with past-life memories have images/memories of violent death, which preoccupies them for some period of time and is likely to cause a PTSD just as persons who are exposed to extremely life-threatening situations develop a PTSD.

What about other psychological characteristics? Tests and qurestionnaire data show that these childen do not confabulate more than other children, are not highly suggestible, do not live in social isolation or in disturbed family relationships, and are apparently not attention-seeking (Haraldsson 1997, 2003). They daydream or get lost in their thoughts more often than other children, perhaps primarily because they are preoccupied with their last-life images.

11.4 Discussion

Any viable interpretation of cases of the reincarnation type has to be able to account for three findings: memories that have been verified as correct in the absence of a normal explanation, birthmarks that fit the wounds of an identified previous personality, and psychological characteristics such as phobias and PTSD. Cases differ greatly and different explanations may apply for different cases. Normal as well as transcendental/paranormal interpretations have been put forward to explain the most impressive cases. The normal explanations have ranged from criticism of the way the investigations have been conducted, namely that they are biased, not sufficiently objective, to failing to consider sufficiently how the cases are culturally molded. Published accounts by Stevenson and those associated with him reveal objective, detailed and thorough the investigations, for example the lengthy reports in Stevenson's four volumes on *Cases of the Reincarnation Type* (Stevenson 1975–1983). They are primarily reports of impressive solved cases whereas the great number of cases that have proved impossible to solve are meagerly dealt with or not all. Stevenson and associates have responded to such criticism (Cook et al. 1983a, b)

Some cases have remained unsolved because they lack verifiable details. Others have verifiable details but were proved wrong (falsified), or are not fitting the life of any potential previous personality that has been identified such as in the above case of Pretiba Gunawardane. The percentage of unsolved cases varies among countries, for example in Sri Lanka they comprise about two-thirds of the cases, which is unusually high, but still below the 80% of cases in the United States.

Cases of the reincarnation type are primarily found in cultures with a overwhelming belief in reincarnation, such as Burma, Thailand, India, Sri Lanka, Lebanon, and Brazil, a Christian country, where over half of the population believes in reincarnation.

However, cases have also been found in the United States (Pasricha et al. 2005; Stevenson 1983; Tucker 2005) and in several countries of Europe (Stevenson 2003). European and United States cases tend to be less impressive and harder to solve than cases found in cultures with high belief in reincarnation. Stevenson's analysis of European cases reveals that their principal characteristics are similar, such as the mean age of first speaking of a previous life, mean age of ceasing to speak spontaneously about it, mentioning the mode of death, and the frequency of violent death.

Birthmarks that correspond to wounds inflicted on an identified previous personality are found in some case, such as in the case of Purnima Ekanayake. It is hard to imagine how cultural influences can cause birthmarks as they are formed in the embryo before the child is born. Stevenson (1997a, b) has written two lengthy volumes about cases with birthmarks and deformities, richly documented with photographs and medical details.

Another criticism concerns the possibility that the correspondence found between statements made by the children of their previous life and events in the life of certain deceased persons may be accidental or due to chance. This is a serious criticism in view of the many cases that remain unsolved. On the contrary, the chance interpretation is made doubtful for the most impressive cases because of the highly specific nature of some of the verified statements made by the children. This interpretation runs into further difficulties when applied to birthmark cases, the phobias and the PTSD, which is found in many children who are not known to have been exposed to life-threatening situations.

Are these children psychically gifted and able to zoom in on facts in the life of a person that lived before they were born? This interpretation runs into difficulty when attempting to explain the psychological aspects and the birthmark cases. And why do these children, zoom back on one particular deceased person, as most show no other psychic abilities?

There are also transcendental interpretations. Possession is sometimes mentioned, also in countries where belief in reincarnation is dominant. Here it is assumed that a deceased person somehow influences a child's behavior and memories.

Another transcendental interpretation is the reincarnation hypothesis, which is commonly accepted in countries with widespread belief in reincarnation. It seems to explain rather easily phobias, PTSD, and the birthmark cases. There is, however, a serious stumbling block. It runs contrary to our present knowledge of memory's dependence on brain functioning. It is hard to see how the reincarnation concept can be accommodated within the current scientific framework without a radical change. That change would have to accept that mind is an independent reality, which could exist without a brain, and which would be the vehicle for the transmission of memory from one brain (which belonged to a deceased person) to another (the child claiming memories).

We have two opposing explanatory models to explain cases of the reincarnation type. On the one side, we have primarily the chance hypothesis, cultural influences and biased reporting. On the other side, we have the transcendental interpretations, primarily reincarnation. The transcendental models seem irreconcilable without a radical change of the present scientific view of the mind–brain relationship.

The evidence supporting the reincarnation hypothesis has been accumulating over the last half of a century. A renowned skeptic like the popular astronomer Carl Sagan wrote: "At the time of writing there are three claims in the (paranormal) field, which in my opinion, deserve serious study" the third being "that young children sometimes report details of a previous life, which upon checking turn out to be accurate and which they could not have known about in any other way than reincarnation" (1996, p. 302).

It has great implications for any theory of the mind–brain relationship if the most impressive cases reveal genuine past-life memories. The present view would have to be revised or extended to incorporate the fact that in rare instances memories/experiences of a person whose brain no longer exists can appear in the brain of a young child. We would have to assume that mind and body are two separate entities, which are only combined for a certain period of time. Furthermore, that, at least for some individuals, some memories that are stored in the mind survive the disintegration of the brain and appear in a new brain that is still developing.

Studies of "reincarnation cases" also indicate that human personality is not only formed by genetics and environmental influences. On the basis of his data, Stevenson (2001) has written at length about the explanatory value of the idea of reincarnation and how it may help us understand some unsolved problems in psychology, biology, and medicine. These include phobias of infancy and early childhood, unusual interests and types of play, unusual aptitudes and skills in early childhood, addictions and cravings in early childhood, gender-identity confusion, differences between one-egg twin pairs, and some birthmarks and birth defects.

References

Cook, E. W., Pasricha, S., Samararatne, G., Maung, U. W., & Stevenson, I. (1983a). A review and analysis of "unsolved" cases of the reincarnation Type: I. Introduction and Illustrative case reports. *The Journal of the American Society for Psychical Research, 77*, 45–62.

Cook, E. W., Pasricha, S., Samararatne, G., Win, M., & Stevenson, I. (1983b). A review and analysis of "unsolved" cases of the reincarnation Type: II. Comparison of features of solved and unsolved cases. *The Journal of the American Society for Psychical Research, 77*, 115–135.

Haraldsson, E. (1991). Children claiming past-life memories: Four cases in Sri Lanka. *Journal of Scientific Exploration, 5*, 233–262.

Haraldsson, E. (1997). Psychological comparison between ordinary children and those who claim previous-life memories. *Journal of Scientific Exploration, 11*, 323–335.

Haraldsson, E. (2000a). Birthmarks and claims of previous life memories I: The case of Purnima Ekanayake. *Journal of the Society for Psychical Research, 64*(858), 16–25.

Haraldsson, E. (2000b). Birthmarks and claims of previous life memories II: The case of Chatura Karunaratne. *Journal of the Society for Psychical Research., 64*(859), 82–92.

Haraldsson, E. (2001). Do some children remember fragments of a previous life? In D. Lorimer (Ed.), *Thinking beyond the brain* (pp. 81–94). London: Floris Books.

Haraldsson, E. (2003). Children who speak of past-life experiences: Is there a psychological explanation? *Psychology and Psychotherapy: Theory Research and Practice, 76*(1), 55–67.

Haraldsson, E., & Abu-Izzeddin, M. (2002). Development of certainty about the correct deceased person in a case of the reincarnation type: The case of Nazih Al-Danaf. *Journal of Scientific Exploration, 16*, 363–380.

Haraldsson, E., & Abu-Izzeddin, M. (2004). Three randomly selected Lebanese cases of children who claim memories of a previous life. *Journal of the Society for Psychical Research., 86*(875), 65–85.

Haraldsson, E., Fowler, F., & Periyannanpillai, V. (2000). Psychological characteristics of children who speak of a previous life: A further field study in Sri Lanka. *Transcultural Psychiatry, 37*, 525–544.

Haraldsson, E., & Samararatne, G. (1999). Children who speak of memories of a previous life as a Buddhist monk: Three new cases. *Journal of the Society for Psychical Research, 63*(857), 268–291.

Mills, A., Haraldsson, E., & Keil, H. H. J. (1994). Replication studies of cases suggestive of reincarnation by three independent investigators. *The Journal of the American Society for Psychical Research, 88*, 207–219.

Pasricha, S. K., Keil, J., Tucker, J. B., & Stevenson, I. (2005). Some bodily malformations attributed to previous lives. *Journal of Scientific Exploration, 19*(3), 359–383.

Sagan, C. (1996). *The demon-haunted world: Science as a candle in the dark*. New York: Random House.

Stevenson, I. (1960). The evidence for survival from claimed memories of former incarnations. *The Journal of the American Society for Psychical Research, 54*(51–71), 95–117.

Stevenson, I. (1975–1983). *Cases of the reincarnation type. Vols. I-IV. Vol. I Ten cases in India* (1975); *Vol. II Ten cases in Sri Lanka* (1977); *Vol. III. Twelve cases in Lebanon and Turkey* (1980); *Vol. IV Twelve cases in Thailand and Burma* (1983). Charlottesville: University Press of Virginia, 1975–85.

Stevenson, I. (1983). American children who claim to remember previous lives. *The Journal of Nervous and Mental Disease, 171*, 742–748.

Stevenson, I. (1997a). *Reincarnation and biology: A contribution to the etiology of birthmarks and birth defects* (Vol. 1 and 2). Westport, CT: Praeger.

Stevenson, I. (1997b). *Where reincarnation and biology intersect*. Westport, Connecticut: Praeger.

Stevenson, I. (2001). *Children who remember previous lives. A question of reincarnation*. Jefferson, NC: McFarland.

Stevenson, I. (2003). *European cases of the reincarnation type*. Jefferson, NC: McFarland.

Stevenson, I., & Haraldsson, E. (2003). The similarity of features of reincarnation type cases over many years: A third study. *Journal of Scientific Exploration, 17*(2), 283–289.

Stevenson, I., & Samararatne, G. (1988). Three new cases of the reincarnation type in Sri Lanka with written records made before verification. *Journal of Scientific Exploration, 2*, 217–238.

Stevenson, I., & Schouten, S. (1998). Does the socio-psychological hypothesis explain cases of the reincarnation type? *The Journal of Nervous and Mental Disease, 186*, 504–506.

Tucker, J. R. (2005). *Life before life*. New York: St. Martin's Press.

Chapter 12
Conclusion

Alexander Moreira-Almeida and Franklin Santana Santos

The empirical data and the arguments presented and discussed in this work represent a serious challenge to the reductionist materialist approaches proposed to explain mental phenomena and their relationship with the brain. The main purpose of this work was to put together arguments and empirical data from a wide variety of perspectives.

Throughout the book the severe limitations of reductionist materialist perspectives were explored. From a historical philosophical perspective, science was distinguished from an ideology (scientism, scientific materialism, physicalism) that involves metaphysical assumptions about the components of the universe. Materialism is not a necessary and logical consequence of scientific research. "Promissory materialism" was exposed as an ancient strategy (dating at least three centuries) that has never fulfilled its hopeful prophecy that it will soon provide a strictly materialistic explanation of the mind.

Two recurrent objections to nonmaterialist perspectives of mind deal with alleged philosophical shortcomings of dualistic perspectives, as well as their incompatibility with physics and a naturalistic explanation of mind and the universe as a whole. Because of this, any theory resembling dualism is often hastily dismissed as naïve or antiscientific. However, in this work we were able to see that the most common objections to dualism have serious problems in the face of careful examination. From the naturalistic perspective, the authors of this book have described nonreductionistic conceptions of mind formulated in terms of the models of modern physics.

Perhaps, the main contribution of this book, in addition to putting together a wide range of perspectives, is to bring to the current debate about the mind–brain

A. Moreira-Almeida, MD, PhD (✉)
School of Medicine, Federal University of Juiz de Fora,
36036-330, Juiz de Fora, MG, Brazil
e-mail: alex.ma@ufjf.edu.br

F.S. Santos, MD, PhD
Institute of Psychiatry, School of Medicine, São Paulo University, 05403-010, São Paulo, Brazil
e-mail: franklin@saudeeducacao.com.br

A. Moreira-Almeida and F.S. Santos (eds.), *Exploring Frontiers of the Mind-Brain*   233
Relationship, Mindfulness in Behavioral Health, DOI 10.1007/978-1-4614-0647-1_12,
© Springer Science+Business Media, LLC 2012

problem a variety of old, widespread, and common human experiences that have been currently neglected or dismissed in the discussion. As in earlier scientific revolutions in other areas of science, we think that this enlarged empirical basis may decisively advance the debate that has been in stalemate for a long time. If a paradigm is proposed to explain mind and its relationship with the body, it must explain the whole range of human experiences, otherwise it would be, at least, incomplete.

The evidence summarized in this book is based on more than one century of investigations, some of them performed by a number of the brightest scientific minds of their times. The available evidence points to the existence of an irreducible and nonmaterial aspect of mind that interacts with and influences material aspects of nature, thereby allowing downward causation.

From an epistemological perspective, ideally, evidence for or against some hypothesis should be based on different kinds of phenomena, collected by different researchers, explored by different research methods, and yet converging to similar conclusions. It is worth stressing that the human experiences discussed in this work are not limited to one sort of phenomena or restricted to a certain culture. In this book, 16 authors presented and discussed data published by dozens of scientists who used different methods to investigate several kinds of experiences that have been reported in many cultures throughout human history. These different types of experiences suggest an immaterial aspect of being that occurs at different ages throughout human lifespan. As soon as children are able to speak, many children begin to report verifiable memories of a previous life. Some other children also start to present mediumistic skills. During adulthood, meditative and mystic states, and several other spiritual phenomena are frequent. Toward the end of life, there are near-death experiences and other end-of-life experiences challenging reductive materialist explanations.

Of course, there is no definitive proof or crucial experiment to confirm the nonmaterial aspects of mind. Scientific development is seldom made of clear cut, discrete steps. Scientific enterprise is a very complex and a much more uncertain endeavor than we would like. As stated by the philosopher of science Imre Lakatos (1970): "*There are no such things as crucial experiments*, at least not if these are meant to be experiments which can *instantly* overthrow a research programme." (p.173).

Any scientific theory passes through an initial period of uncertainty, incompleteness, and suspicion; however, if the theory and its proponents are good enough, these shortcomings are overcame (Lakatos 1970). A good example is the history of evolution by natural selection, although it is now a very well established and funded paradigm in biology, this has not always been the case. Many important supportive evidence and theoretical developments came only decades later, in the first half of twentieth century, through the integration of genetics with Darwinian evolution in what is known as "modern synthesis" or "evolutionary genetics" (Wade 2008).

Charles Darwin, himself, admitted that it was not possible to present, at his time, a direct and compelling proof of natural selection. He empathized with those who did not become convinced of his theory:

I am not at all surprised that you are not willing to admit natural selection: the subject hardly admits of direct proof or evidence. It will be believed in only by those who think that it connects & partly explains several large classes of facts: in the same way opticians admit the undulatory theory of light, though no one can prove the existence of ether or its undulations. (Darwin 2010a, letter no. 3608).

The main advantage of evolution by natural selection over competing theories was its explanatory power, which connected and made sense of a large body of empirical evidence that were not well accommodated by competing paradigms:

> In fact the belief in natural selection must at present be grounded entirely on general considerations. (1) on its being a vera causa, from the struggle for existence; & the certain geological fact that species do somehow change (2) from the analogy of change under domestication by man's selection. (3) & chiefly from this view connecting under an intelligible point of view a host of facts. (Darwin 2010b, letter no. 4176).

The empirical data revised in this book is hard to reconcile with reductionist materialist perspectives of mind. One alternative mind–brain theory that seems to fit better with the presented data is the William James' transmission theory of cerebral action, in which "brain *can* be an organ for limiting and determining to a certain form a consciousness elsewhere produced" (James 1898: 294). Also according to James, the transmission theory (in which brain has a permissive or transmissive function) has an explanatory advantage over the production theory (in which brain produces mind). In addition to explaining the predictions of the production theory (specifically, the concomitance of changes in brain and consciousness, i.e., brain correlates of mental phenomena):

> The transmission theory also puts itself in touch with a whole class of experiences that are with difficulty explained by the production theory. I refer to those obscure and exceptional phenomena reported at all times throughout human history, which the 'psychical researchers' (…) are doing so much to rehabilitate; such phenomena, namely (…) premonitions, apparitions at time of death, clairvoyant visions or impressions, and the whole range of mediumistic capacities (…). All such experiences, quite paradoxical and meaningless on the production theory, fall very naturally into place on the other theory. We need only to suppose the continuity of our consciousness (…) (James 1898: 298–299).

As recently stated by David Chalmers (2003), several important limitations of materialistic perspectives have been exposed and the main objections to nonmaterialistic perspectives have been overcome, so nonmaterialistic perspectives should now receive further consideration. He suggested that the two best candidates for a scientific theory of mind were interactionist dualism and panprotopsychism. He stated that, contrary to his initial inclination, he became a kind of dualist because of his scientific attitude of "acknowledge (not dismiss) the data" (Chalmers, 2009:4). We agree with his conclusion about the place of consciousness in nature:

> It is often held that even though it is hard to see how materialism could be true, materialism must be true, since the alternatives are unacceptable. As I see it, there are at least three prima facie acceptable alternatives to materialism on the table [interactionist dualism, epiphenomenalism, and panprotopsychism], each of which is compatible with a broadly naturalistic (even if not materialistic) worldview, and none of which has fatal problems. So given the clear arguments against materialism, it seems to me that we should at least tentatively embrace the conclusion that one of these views is correct. Of course all of the views (…) need to be developed in much more detail, and examined in light of all relevant scientific and philosophical developments, in order to be comprehensively assessed. But as things stand, I think that we have good reason to suppose that consciousness has a fundamental place in nature. (Chalmers 2003:41–42).

Undoubtedly, there is still the need of much more empirical research and theoretical development of nonreductionist views of mind. Given the evidence available and our limited understanding of mind, nonmaterialistic perspectives deserve at least the same opportunity of development as the materialistic ones. Intellectual freedom is needed to develop and improve paradigm candidates without suppression by dogmatism and intolerance. Also, according to Lakatos (1970), scientific development happens in a kind of Darwinian selection of competing paradigms candidates where the fittest survives:

> It would be wrong to assume that one must stay with a research programme until it has exhausted all its heuristic power, that one must not introduce a rival programme before everybody agrees that the point of degenerations has probably been reached. (...)
> The history of science has been and should be a history of competing research programmes: the sooner competition starts, the better for progress. 'Theoretical pluralism' is better than 'theoretical monism' (155).

Exactly because of this, we live in an exciting time. If we humbly recognize our very limited knowledge about consciousness, and, at the same time, we boldly, rigorously, and creatively face the mind–brain problem, we, as human beings, may march to a deeper understanding of our own nature. This is undoubtedly a tough and challenging enterprise, however it is definitely a path worthwhile trailing:

> I believe that it would be worth trying to learn something about the world even if in trying to do so we should merely learn that we do not know much. This state of learned ignorance might be a help in many of our troubles. It might be well for all of us to remember that, while differing widely in the various little bits we know, in our infinite ignorance, we are all equal. (Popper 1995:29).

References

Chalmers, D. J. (2003). Consciousness and its place in nature. In S. Stich & F. Warfield (Eds.), *Blackwell guide to philosophy of mind*. Oxford: Blackwell.

Chalmers, D. J. (2009). Mind and consciousness: Five questions. In P. Grim (Ed.), *Mind and consciousness: Five questions*. London: Automatic Press.

Darwin, C. (2010a). *Darwin Correspondence Project Database*. Retrieved May 25, 2010, from http://www.darwinproject.ac.uk/entry-3608/ (letter no. 3608).

Darwin, C. (2010b). *Darwin Correspondence Project Database*. Retrieved May 25, 2010, from http://www.darwinproject.ac.uk/entry-4176/ (letter no. 4176).

James, W. (1898/1960). Human Immortality: Two supposed objections to the doctrine. In G. Murphy, R. O. Ballou (Ed.), *William James on psychical research* (pp. 279–308). New York, Viking Press.

Lakatos, I. (1970). Falsification and the methodology of scientific research programmes. In I. Lakatos & A. Musgrave (Eds.), *Criticism and the growth of knowledge* (pp. 91–195). Cambridge: Cambridge University Press.

Popper, K. R. (1995). *Conjectures and refutations – the growth of scientific knowledge*. London: Routledge.

Wade, M. (2008). Evolutionary genetics. Retrieved Sept 16, 2011. In N. Edward Zalta (Ed.), *The Stanford Encyclopedia of Philosophy* (Fall 2008 Edition). Retrieved sept 16, 2011, from http://plato.stanford.edu/archives/fall2008/entries/evolutionary-genetics.

About the Editors

Alexander Moreira-Almeida, M.D., Ph.D. was trained in psychiatry and cognitive-behavioral therapy at the Institute of Psychiatry of the University of São Paulo, Brazil, where he also obtained his Ph.D. in Health Sciences investigating the mental health of Spiritist mediums. Formerly a postdoctoral fellow in religion and health at Duke University, he is now Professor of Psychiatry at the Federal University of Juiz de Fora School of Medicine and Founder and Director of the Research Center in Spirituality and Health, Brazil (www.ufjf.br/nupes-eng). His main research interest involves empirical studies of spiritual experiences as well as the methodology and epistemology of this research field. His publications are available at www.hoje.org.br/elsh.

Franklin Santana Santos, M.D., Ph.D. was trained in geriatrics at the Clinical Hospital of the University of São Paulo, Brazil, where he also obtained his Ph.D. in Health Sciences investigating delirium in elderly patients. Formerly a postdoctoral fellow in cognitive disturbances at Karolinska Institute (Sweden), he is now Professor of postgraduate program of University of São Paulo School of Medicine and collaborator researcher of Laboratory of Neuroscience (LIM-27) at Institute of Psychiatry of the University of São Paulo, Brazil. He is a leader of the studies in issues related to death, dying, and palliative care in Brazil. His main research interest involves cognitive disturbances, thanatology, palliative care, and medical education, topics about which he has published several articles and books.

Index

Printed by Publishers' Graphics LLC USA
MRO20120404-10
2012